GLORY, GRIT AND GREATNESS

GLORY, GRIT AND GREATNESS

GREAT AMERICANS FADING FROM OUR MEMORY

STEPHEN CARR

IZZARD INK PUBLISHING

IZZARD INK PUBLISHING
www.izzardink.com

Copyright © 2025 by Stephen Carr

All rights reserved. Except as permitted under the U.S. Copyright Act of 1976, no part of this publication may be reproduced, distributed, or transmitted in any form or by any means, or stored in a database or retrieval system, electronically or otherwise, or by use of a technology or retrieval system now known or to be invented, without the prior written permission of the author and publisher.

Library of Congress Cataloging-in-Publication Data
Names: Carr, Stephen L. author
Title: Glory, grit and greatness : great Americans fading from our memory / Stephen Carr.
Description: Salt Lake City : Izzard Ink Publishing, 2025.
Identifiers: LCCN 2025006828 (print) | LCCN 2025006829 (ebook) |
ISBN 9781642281231 hardback | ISBN 9781642281224 paperback |
ISBN 9781642281200 ebook
Subjects: LCSH: United States—Biography | National characteristics, American | United States—History | LCGFT: Biographies
Classification: LCC E176 .C29 2025 (print) | LCC E176 (ebook) | DDC 973.09/9—dc23/eng/20250501
LC record available at https://lccn.loc.gov/2025006828
LC ebook record available at https://lccn.loc.gov/2025006829

Designed by Daniel Lagin
Cover Design by Andrea Ho

First Edition

Contact the author at info@izzardink.com

eBook ISBN: 978-1-64228-120-0
Paperback ISBN: 978-1-64228-122-4
Hardback ISBN: 978-1-64228-123-1
Audiobook ISBN: 978-1-64228-124-8

To

Leslie Ajac Deese

When she smiled I saw sunflowers in her eyes.

So it's home again, and home again, America for me!
My heart is turning home again and there I long to be
In the land of youth and freedom beyond the ocean bars,
Where the air is full of sunlight and the flag is full of stars.

—HENRY VAN DYKE

CONTENTS

PREFACE   xi

1. **IF THERE IS ONLY ONE PLANE LEFT**   1
2. **THE MOST HATED MAN IN AMERICA**   35
3. **RAYS OF GLORY**   59
4. **THE PERFECT STICK**   77
5. **SUZIE Q**   107
6. **HEROES OF THE LAKES**   135
7. **GOD BLESS AMERICA**   163
8. **THE YEGG HUNTERS**   185
9. **I AIN'T NO MUSEUM PIECE**   217

NOTES   241

PREFACE

It was Memorial Day weekend, about fifteen years ago I think, a beautiful late-spring morning in San Diego, California, sunny and clear, mid-seventies, with a light breeze blowing in from San Diego Bay a mile and a half away.

I was walking up a shallow grass slope on my way to the recreation center in Balboa Park when I heard a bell ringing behind me, a low, rhythmic chime that sounded like a grandfather clock slowly tolling midnight.

Across a lawn at the bottom of the slope, two elderly men stood on a stage outside a veterans museum, one of whom held the lanyard of a brass bell dangling from the end of a tall, cane-shaped stand, the other of whom was reading off names from a military roster. After each name, the bell rang once. Clearly, the ceremony was meant to honor American war dead, though in which war or wars I don't know.

Spread out below the stage were a dozen-odd folding chairs, all but a few of which were empty. After watching from the slope for a

PREFACE

moment, I continued on to the rec center, feeling sheepish for not joining the people sitting in the chairs. I thought it a shame that almost no one, myself included, took the time on that beautiful morning to sit and watch the ceremony and listen to the names of heroes killed in action.

The incident reminded me of something I'd been noticing for a long time, especially while watching videos of pop history quizzes on the internet. You may have seen them. A reporter approaches members of the public, often young people, and tests their knowledge of noteworthy figures from American history—Founding Fathers, military leaders, presidents, etc. More often than not, the responses are woeful, one blank stare or wild guess after another.

Watching these videos made me wonder what kids were being taught about American history, so I thumbed through some high school history textbooks published in the last twenty years or so. Many people I'd learned about in school in the 1960s and 1970s were either left out or glossed over.

I believe America's past is slipping away from us, that many of our greatest Americans, well known in their day, are no longer getting the recognition they deserve, are fading from our national memory. This book is about some of those Americans.

In a country with a history as rich as ours, one of the biggest challenges was deciding whom to include. I wanted to avoid the obvious. There are no towering statesmen, leaders of national crusades, or great humanitarians portrayed here, no Daniel Websters, William Lloyd Garrisons, or Clara Bartons.

I chose people from many eras and walks of life. Most rose from humble roots. Some were immigrants or first-generation Americans. Some were high school dropouts. All of them, through sheer grit or genius, or both, did extraordinary things.

PREFACE

Featured are a U.S. president, an industrialist, an entertainer, an athlete, some lawmen, a songwriter, and heroes from three wars.

The entertainer was a woman. The rest of the main subjects were men, but they tended to attract formidable women, whom I also write about.

The people in the book were a remarkable group. They transformed our way of life, inspired us, made us cheer, kept our streets safe, and fought for our freedoms and sometimes died for them. They deserve to be remembered.

That's not to say they were perfect. Far from it. Like the rest of us, like America itself, they had their share of flaws. The president was an oddball, the songwriter a temperamental workaholic. The entertainer had an eating disorder. The tycoon was a cutthroat businessman. The lawmen shot first and asked questions later, and the athlete was a philanderer. I believe their greatness shines through their faults.

As for the war heroes, I chose to focus on their valor and sacrifice rather than whatever flaws they may have had. It's hard to find fault with those who put their lives on the line defending our country, especially in battles that saved America, as some of the heroes portrayed here did.

The book is for people like me who are fed up with America bashing, who would rather salute our flag than sully it, who aren't ashamed of the pride they feel in our nation's glorious past.

History buffs will have heard of many of the people in the book. When I first put pen to paper, I barely knew them. Since then, I've come to realize how truly remarkable they were. I've also learned that behind every great American is a terrific story.

I hope you agree.

CHAPTER 1

IF THERE IS ONLY ONE PLANE LEFT

The Torpedo Plane Squadrons at the Battle of Midway

It was almost midmorning before the order came blaring over the ready-room loudspeaker—*Pilots and crews man your planes.*

The men of the carrier USS *Yorktown*'s torpedo-plane squadron cast aside their paperbacks and half-written letters, stubbed out their cigarettes after a quick final drag, and scrambled up to the fantail, where a dozen single-engine Douglas Devastators, packed wing-to-wing, their propellors rotating, their engines idling, a thirteen-foot, cigar-shaped torpedo hugging their bellies, were already cranked up and ready to go.

The two-man crews—one pilot and one gunner per aircraft—climbed aboard the planes, which one by one roared down the carrier's flight deck and took wing over the Pacific Ocean.

Rumbling skyward on that sunny Thursday, amid a light breeze and scattered clouds, "good hunting weather" as the saying went, the Devastators formed up with Dauntless dive bombers and Wildcat fighters from the *Yorktown*, three dozen planes in all, and headed out in search of the vaunted Kidō Butai, the Japanese fleet that had just attacked the American base on Midway Atoll.

Sitting in the Devastator's gun turret behind the cockpit, Petty Officer Third Class Lloyd Childers had a bird's-eye view of the sprawling American task force as it faded from sight until nothing remained but miles of empty sea. The enemy fleet lay far beyond the horizon, a good hour's flight away, so he sat back, scanned the sky for bogeys, and waited.

The biscuit salesman's son from Oklahoma had wanted to be a United States Marine until a couple of arrogant sergeants at the recruiting depot rubbed him the wrong way. He'd walked out, joined the navy, and never looked back.

Compared to the hard times he'd faced as a boy—growing up poor in Oklahoma City, being shunted between two households after his parents' divorce, working long hours from the age of ten to help support his mom and younger brothers—he found navy life a pleasure, especially aboard the *Yorktown*, a huge floating city with roomy bunks, plentiful chow, and best of all, a soda fountain that served soft drinks and homemade ice cream. He loved ice cream.

A regular eager beaver, always willing to volunteer, he'd trained hard to become a radio operator and had just been promoted to petty officer. He dreamed of attending the U.S. Naval Academy and becoming an aviator and had been working with a math tutor to prepare for the entrance exam. Meanwhile, well aware of the risks, he'd volunteered to be a gunner aboard a torpedo plane.

Today, he was traveling a little lighter than usual, bringing along only the bare essentials: the parachute strapped to his back, the bright yellow "Mae West" life jacket draped over his shoulders, his dog tags in the pocket of his flight jacket, and his pistol. He'd left his watch and wallet behind in his locker. They would make nice keepsakes for his family. Besides, where he was going he wouldn't need them.

IF THERE IS ONLY ONE PLANE LEFT

He was also going aloft with a hearty breakfast in him. He and his fellow gunners had been treated to steak and eggs, fare usually reserved for the officers. The reason for the special chow was no secret. "Last meal for the condemned men," someone in the chow line blurted, triggering nervous laughter.

Later, just before taking off, the slender Okie with the gentle brown eyes, dimpled smile, and prairie drawl tried to break the tension with a quip of his own. "Today, theoretically, I'm a man," he was heard to say. "Let's celebrate."

It was June 4, 1942. Lloyd Childers' twenty-first birthday.

Aboard the *Yorktown*'s sister ship, the carrier USS *Enterprise*, Lieutenant John Thomas Eversole knew what fate awaited him.

The tall, handsome twenty-seven-year-old from Pocatello, Idaho, U.S. Naval Academy class of '38, had already survived a brush with death. Returning from a scouting mission, he'd been forced to ditch his Douglas Devastator in the ocean in a thick fog in the middle of nowhere. He'd been lucky that day (a rescue mission had found him), but now, on the eve of battle, he sensed his luck had run out.

"He knew he didn't have any real chance of surviving," his best friend, dive-bomber pilot Norman "Dusty" Kleiss, remembered seventy years later. "Tom was willing to trade his life for his country."

The problem for Tom Eversole and the other American torpedo-plane pilots was their outmoded planes. Derided as "flying coffins," the Devastators—or "TBDs" (Torpedo Bomber Douglas), as they were often called—were the butt of a running joke that they never devastated anything except the men who flew in them. When introduced five years earlier as the Navy's first all-metal monoplane, the TBDs had been state-of-the-art, but by mid-1942 they were obsolete. Although sturdily built, they were armed with only a single .30 caliber machine gun, mocked as a "pea shooter." Worse, the planes were too slow for their hazardous mission—fatally slow, as events would soon prove.

Kleiss saw how slow when he and Eversole flew patrols together, Kleiss in his Dauntless dive-bomber and Eversole in his TBD: "He'd fly at full throttle at 90 knots or so [103 miles an hour] and I would stall at that speed, so I'd have to fly S-turns above him to allow him to keep up."

In battle, lumbering in just over the wavetops, to drop a torpedo designed to strike a ship below the waterline, the TBDs were clay pigeons for shipboard antiaircraft fire and swift, nimble Zeros, Japanese fighter planes. The U.S. Navy estimated that as many as four out of five TBDs would be shot down even before they had a chance to launch their torpedoes.

Chilling as it was, the estimate was optimistic. Ideally, the TBD squadrons were supposed to be part of a coordinated attack that included fighters and dive bombers to help distract the antiaircraft gunners and Zeros, but if the plan went awry and the vulnerable torpedo planes ended up going in alone, the odds against them would be even worse. The attack could become an outright massacre.

On the morning of June 4th, as the *Enterprise* prepared to launch its planes to seek out and attack the Kidō Butai, the two friends met up to shake hands and wish each other luck. For Dusty Kleiss, the moment would forever stir deep emotions. "I looked for Tom so I could say goodbye," he would recall, choking back tears. "We both knew his odds of returning were very low, that he was on a suicide mission for his country. It was hard to walk away."

At dawn, jarred awake by gunfire, fifteen pilots in skivvies tumbled out of bed into the semidarkness wondering who the hell was shooting up their barracks.

IF THERE IS ONLY ONE PLANE LEFT

No, it wasn't a Japanese attack. The nearest enemy combatant was half an ocean away. The shooter was none other than the pilots' skipper, Lieutenant Commander John Waldron, U.S. Naval Academy class of 1924, commander of the carrier USS *Hornet*'s torpedo-plane squadron. Standing in the middle of the barracks, he'd just emptied his Colt .45 through the open entryway into a canebrake.

It was May 28th at Ewa Field on the Island of Oahu, Hawaii, and in just one week the squadron would be flying into battle for the first time as part of the impending showdown with the Japanese fleet in the central Pacific around Midway Atoll. Next time the shooting would be for real.

For months, the hard-charging forty-one-year-old from South Dakota had been riding herd on his squadron of mostly green young ensigns fresh out of flight school, feverishly trying to ready them for combat, and he wasn't about to ease up now. As soon as the gunfire died down, he began bellowing orders: "Hit the deck everybody. This is no drill. This is it. Rise and shine. We have work to do."

Someone once quipped that Waldron was born with six-shooters in his hands. One look at him and you knew why. Tall, lean, and leathery, with deep-set eyes that, when rankled, lashed out like a bullwhip, he had the look of a gunslinger. He didn't usually have to fire his .45 to get your attention. One stern look was enough.

Raised in the wilds of Saskatchewan, Canada, where his family had moved when he was a boy, one-eighth Oglala Sioux, a heritage he took great pride in, he'd been mocked at the Naval Academy as "the class Redskin" and a "seagoing cowpuncher." He couldn't have cared less. Like a lot of maverick officers, he gravitated toward the Navy's budding aviation branch, where he thrived. By the time of the Midway campaign, he was among the most respected squadron leaders in the fleet. What stood out most was his hell-bent determination.

Petty Officer Lloyd Childers
(COURTESY OF STEVEN CHILDERS.)

Lt. (j.g.) John Thomas Eversole

Lieutenant Commander John Waldron.

IF THERE IS ONLY ONE PLANE LEFT

"If we run out of gas," he once growled, "we'll piss in the tanks." He chose a single word for his squadron's motto—*Attack!*

Waldron had waged a furious campaign to get his squadron equipped with the new Avenger torpedo planes, which were just coming off the assembly line and were faster and better armed than the TBDs. After months of promises and delays, the new planes would arrive in Hawaii too late, one day after the *Hornet* had already sailed for Midway.

Like their colleagues aboard *Yorktown* and *Enterprise*, Waldron's men would be attacking in the antiquated TBD's, all the more reason, in his mind, for intense training. "Before we are done here," he warned his men, "you are going to wish every plane was in hell and I was down there with 'em."

He honed his pilots to a fine edge with eight hours a day of flying, daily physical exercise, and countless drills and chalk talks at all hours on everything from tactics to the TBD's hydraulic system. Even the pilots who chafed under the grueling regimen, and groused about "the old goat" behind his back, couldn't help admiring him. His élan was infectious. Under his leadership, the squadron became a band of brothers, ready to follow him anywhere.

For all its intensity, the training hadn't been ideal. Due to wartime shortages, most of Waldron's pilots would fly into battle without ever having launched a torpedo. Much of their combat training had consisted of flat-hatting around the countryside near the U.S. naval base at Norfolk, Virginia, "attacking" tree groves that stood in for the Japanese fleet. When the time came for combat, what the young pilots lacked in experience they would try to make up for it with esprit and raw courage.

On June 3rd, the night before the battle, Waldron typed up a message to the squadron that closed with the words: "I want each of us to do his utmost to destroy our enemies. If there is only one plane left to make a final run in, I want that man to go in and get a hit. May God be with us all. Good luck, happy landings, and give 'em hell."

He and his men were determined to do exactly that, or die trying.

The Japanese would be fighting for the greater glory of their emperor. The Americans would be fighting for revenge, something they'd been after since December 7, 1941—the date which would "live in infamy"—the day of Japan's sneak attack on the U.S. Pacific Fleet anchored at Pearl Harbor on Oahu.

Taking off from carriers two hundred miles offshore, waves of crack Japanese pilots, flying Nakajima torpedo planes, Aichi dive-bombers, and Zeros, had sunk or crippled five battleships—the pride of the U.S. Navy moored along Battleship Row—had blasted to pieces hundreds of warplanes packed wing-to-wing down the center of the island's airfields, and had killed two thousand Americans.

Newsreel footage of the battleship *Arizona* bombed to a twisted hulk, billowing black smoke, was seared into the nation's memory. "Remember Pearl Harbor" became America's battle cry.

The worst was yet to come. Over the next several months one hammer blow after another left the U.S. and its ally Great Britain reeling. Singapore, Great Britain's "Gibraltar of the East," surrendered to the Japanese in mid-February, in what Winston Churchill called "the worst disaster" in British history. South of Manila in the Philippines, America's island fortress of Corregidor held out for four agonizing months, before finally surrendering in early May.

IF THERE IS ONLY ONE PLANE LEFT

Top: A Douglas Devastator with its .30 caliber "pea shooter" rear of center; *Bottom:* TBD's being readied for flight with torpedoes under their bellies.

By then, Dai Nippon—"Great Japan"—had cast its shadow over nearly one-seventh of the globe, including much of East Asia and the western half of the Pacific Ocean. Backed by the world's most powerful navy, Japan's conquests stretched from the Kuril Islands off the coast of Siberia to the Coral Sea off the coast of Australia, from the gates of India to the far-flung islands of Micronesia.

GLORY, GRIT AND GREATNESS

Twelve hundred miles west of Oahu, halfway between Japan and California, tiny Midway Atoll, two sand-covered islands less than two miles long tucked inside a coral lagoon, home to little else but gooney birds and an airstrip, had suddenly become a main U.S. base, the front line of America's defenses in the central Pacific.

Speculation ran wild about where the Japanese would strike next. India, Australia, even Siberia were all potential targets. The brilliant, soft-spoken head of Japan's navy, Admiral Isoroku Yamamoto, a born gambler, whose passions were naval warfare and all-night poker games, wanted to invade Hawaii and use it as a springboard for attacking America's west coast.

Standing in his way were the remnants of the U.S. Pacific Fleet: a couple of battleships, a handful of cruisers and destroyers, and most menacingly, three aircraft carriers—*Yorktown*, *Enterprise*, and *Hornet*— that had escaped destruction at Pearl Harbor. As long as these carriers remained afloat, Yamamoto couldn't rest easy. He planned to lure them into battle using Midway as bait. The Japanese would invade the atoll, then pounce on the U.S. fleet when it came to the rescue, a mismatch he was sure he would win. Victory, to quote a Japanese saying, would be "as easy as twisting a baby's arm." The U.S. Navy would be vanquished, the carriers destroyed. Japan would be the master of the Pacific Ocean, its fleet free to roam unchallenged.

Midway would be a new kind of sea battle for a new age—the age of flight. With the advent of aircraft carriers—whose torpedo planes and dive bombers could reach an enemy fleet with ship-killing ordinance from more than 150 miles away—battles could now be fought over enormous distances, between opposing fleets that never sighted each other. The key to carrier warfare was to hit first and hit hard, find the enemy before he found you and crush him.

IF THERE IS ONLY ONE PLANE LEFT

Yamato amassed over a hundred warships for the Midway operation, so many his main worry was that the Americans would refuse to come out and fight. To make his command less intimidating, he split it up. One task force would be diverted north to attack the Aleutian Islands in Alaska. Another would cover the landing of five thousand troops on Midway's beaches. Yet another, including Yamamoto himself, would trail several hundred miles behind as a reserve.

The tip of the spear was a fourth task force, the Kidō Butai ("Strike Force"), which included all four of Yamamoto's big carriers and their two-hundred-plus warplanes. Their mission was to pulverize Midway's defenses, paving the way for the beach landing, then sink the U.S. carriers after they steamed out of Pearl Harbor and came to the rescue.

"Our hearts burn with the conviction of a sure victory," a Japanese commander wrote in his diary. The Japanese had reason to feel confident. The Kidō Butai had four carriers to the Americans' three, its planes were generally superior to those of the Americans, and its pilots were a battle-savvy elite, the same group that had decimated the U.S. fleet at Pearl Harbor.

But unbeknownst to the brilliant poker player, America had an ace up its sleeve. An equalizer. The U.S. Navy had broken the Japanese code and had been alerted to the Midway attack. Yamamoto was sailing into an ambush.

Even so, everything would depend on the skill and courage of the men who rode the planes into danger, including the eighty-two pilots and gunners manning the forty-one TBDs in the U.S. fleet's three torpedo-plane squadrons. Too many names to list let alone remember.

The pilots were almost all officers, mostly lieutenants and ensigns in their mid-to-late twenties, college graduates from rich or upper-middle-class families. Many were accomplished athletes. Ensign Grant Teats (a bruiser who brooked no nonsense about his last name) had been a track star at Oregon State. Ensign Glenn Hodges had played basketball for the Georgia Bulldogs. Lieutenant James Owens had played quarterback at U.S.C.

The gunners were enlisted men mostly in their early-to-mid-twenties, high school graduates from working-class families. What accolades there were after the battle went mostly to the pilots, all of whom would be awarded the Navy Cross. The gunners would receive a lesser medal, the Distinguished Flying Cross, and most would be quickly forgotten.

Like John Waldron, the other two squadron leaders who would lead these men into battle were tough-minded Naval Academy grads.

Thirty-seven-year-old Lieutenant Commander Eugene Elbert Lindsey from Sprague, Washington, Annapolis class of 1927, had been a chronic screw-up at the Academy, constantly in trouble for his slovenly appearance and shirking his duty. ("His brow is unfurrowed by worry," is how a classmate put it.) In his final year, Lindsey was adjudged "wholly unsuited" to be a naval officer and nearly expelled.

After graduating, he married the daughter of a surgeon, began raising a family, and cleaned up his act. A photo taken about the time he finished flight school of a tight-lipped lieutenant with pockmarked skin and thick black hair who glared at the camera, revealed an intensity at odds with his lax attitude at the Academy. By mid-1940 he was respected enough to be given command of the TBD squadron assigned to the *Enterprise*.

The Battle of Midway would test his grit as well as his valor. A few days prior, he cracked up his TBD as he was making a carrier landing, banging up his face and ribs (the TBD didn't come equipped with a shoulder harness) and wrenching his back so badly the

Clockwise from upper left: Lieutenant Commander Eugene Lindsey; Lieutenant Commander Lem Massey; Admiral Chūichi Nagumo; Admiral Isoroku Yamamoto.

doctors thought it might be broken. On June 4th, he would be in no condition to even squeeze into a cockpit, let alone pilot an aircraft.

Fun-loving, moon-faced Lieutenant Commander Lance "Lem" Massey from upstate New York, Annapolis class of 1930, was, at age thirty-two, one of the fleet's youngest squadron commanders, and among the most laid back. While other senior officers affected a look of steely resolve while posing in their uniforms, pictures of Massey showed him grinning like a Cheshire Cat, his cap pushed rakishly to the side over eyes alight with mirth. In February 1942, as an assistant squadron commander, he'd led the war's first airborne torpedo attack that sank a Japanese ship, earning him the Distinguished Flying Cross and command of the *Yorktown*'s TBD squadron.

Waldron, Lindsey, and Massey were family men, all married with children. Many of their pilots were red-blooded bachelors, young hotshots who, when they weren't at sea or in training, drank hard and partied hard, never missing a chance to head into Norfolk to troll the watering holes for young, hot, equally red-blooded navy nurses. Dashing young combat pilots were almost as alluring as movie stars, and those who wanted to scored almost at will. Hookups became a routine part of life in the BOQ (bachelor officers' quarters). On weekend mornings, the place was full of women showering and having breakfast after an all-night romp. In sick bay, venereal disease carried less of a stigma than a bad case of sunburn. No one begrudged the pilots a few conquests before they flew off to war.

On a sleeper train, Ensign George Gay encountered a comely young woman who was occupying the berth below his. Some friendly conversation led to a pleasant dinner and nightcap in the dining car that ended with a polite parting of the ways. When it came time for Gay to turn in, he saw the woman in bed in the lower birth. She didn't say a word, just smiled and folded back the covers, and Gay slid in beside her.

The dark-eyed twenty-five-year-old Texan with the matinee-idol looks—he bore a passing resemblance to a young Clark Gable—had dreamed of being an aviator ever since taking a ride in an airplane at the Dallas State Fair when he was a boy. After two years of college, with war clouds gathering, he dropped out to join the Army Air Corps. "My mind is in the air, my heart is in the air, so why not pitch the old epidermis into the air too?" he chirped to his dad. Unfortunately, the old epidermis failed him. He flunked the physical. Undaunted, he spent time on a construction crew building bridges to get in shape. He was thinking about going to England to join the R.A.F. when a friend suggested he become a U.S. Navy pilot. Gay was incredulous. The Navy had pilots? Still, he gave the idea a try. Being accepted into the Navy's cadet aviator program was the happiest moment of his life.

After he joined the *Hornet*'s TBD squadron, reality set in. He realized he would be flying into battle in an outmoded plane having never before carried a torpedo in flight. "Sometimes I wonder what I am doing way out here," he confided to his diary. He brooded about facing combat for the first time, picturing its horrors so vividly he swore he could hear them and even smell them. He feared he would crack under fire.

For Gay and the other TBD airmen, when the U.S. fleet set sail for Midway at the end of May, the nightly pleasures ashore gave way to boredom and mounting tension aboard ship. With little to do and the fighting nearly a week away, men killed time whatever way they could: chain-smoking, playing poker, reading. Men kept awake at night by the gnawing pressure catnapped during the day. Amid the idleness, rumors spread like brush fires. Scuttlebutt had it that Yamamoto had slipped past the U.S. fleet and was about to launch a sneak attack on Hawaii. Or maybe California.

Aboard the *Hornet*, Ensign Rusty Kenyon, whose wife Brownie was expecting their first child, broke the tension with limericks, one of which found its way into the history books:

> There was a young girl from Madras
> Who had a beautiful ass,
> Not rounded and pink,
> As you probably think,
> But with long ears, a tail, and eats grass.

Anyway, it was good for a chuckle. There was also music. Between games of acey-deucey, gunner Frank Polston, a Missouri farm boy, entertained his buddies by strumming the song "I've Been Working on the Railroad" on his ukulele.

On the *Enterprise*, gunner John Lane soothed his nerves listening to recordings of soprano Miliza Korjus, "The Berlin Nightingale." Like Tom Eversole, the pilot he flew with, Lane couldn't shake the feeling his number was up.

On Wednesday, June 3rd, at half past nine in the morning, the boredom ended. As the three U.S. carriers steamed in slow circles 350 miles north of Midway, an urgent dispatch arrived from Pearl Harbor. Although the source of the information—the U.S. Navy code breakers—was a well-guarded secret, the message itself went out to the entire fleet: the Japanese carriers would hit the atoll just after dawn.

So this was finally it. The battle for mastery of the Pacific would happen the next morning, June 4th. For the TBD crews, after nearly a week cooped up aboard ship, the news was almost welcome. They spent Wednesday going over their planes with a fine-toothed comb,

IF THERE IS ONLY ONE PLANE LEFT

checking and rechecking every bolt and bullet. That night, the prevailing mood was grim determination. No one knew which side would win the battle, but there was little doubt that tomorrow, barring a miracle, a lot of aviators were going to die.

Lem Massey broke out a fifth of Scotch he'd smuggled into his cabin and shared it with a couple of shipmates. The skipper of the *Yorktown*'s TBD squadron was a decorated combat veteran, as tough and brave as anyone in the fleet, a man with a friendly, upbeat personality, hardly the despairing type. But tonight he couldn't help it. As he sipped his Scotch, he bared his deepest fears. Maybe he shouldn't be talking like this, he said, but he didn't see how he and his men could possibly survive the next day. The odds were too long. He was convinced they would all be wiped out.

Some men wrote farewell letters to their loved ones. "Honey, if anything happens to me . . . don't become an old maid," gunner Otway Creasy, Jr., a machinist's son from Virginia, told his wife Robbie. "Find someone else and make you a happy home. Don't be worrying about me."

Ensign Grant Teats, the track star from Oregon, sounded matter-of-fact, almost businesslike, in a letter to his folks. He informed them that he'd paid his taxes and all his bills and boxed up his civilian clothes. He was sending them a check to "help out with incidentals." He closed on a wistful note:

> I guess this is about all. If I had a drink now it would be a toast to the Japanese navy. Bottoms up. Write soon.
>
> Love, Grant

Tallyho. Hawks at angels twelve.

When the alert was broadcast at six thirty in the morning, every eye on Midway became riveted on the brightening western sky.

There they were, right on schedule, soaring high in the air, like raptors stalking their prey. A hundred Japanese warplanes were approaching the atoll at 12,000 feet.

They shot up a group of fighters sent from Midway's airstrip to intercept them, and within half an hour turned the atoll into a mass of burning wreckage. Fuel storage tanks, water and gas lines, an ammo dump, a power plant, a medical dispensary, even the post office, were demolished. Yet for all that, the destruction wasn't complete. The commander of the Japanese squadrons radioed back to the Kidō Butai recommending a second airstrike.

One hundred fifty miles northwest of the atoll on his flagship, the carrier *Akagi*, Admiral Chūichi Nagumo, the Kidō Butai's commander, agreed. He'd been holding half his planes in reserve on his carriers' hangar decks, ready to attack the U.S. carriers if they showed up sooner than expected. But that precaution no longer seemed necessary. Japanese search planes had been scouring the sea since first light and had seen nothing. Certain the U.S. fleet was nowhere near, Nagumo ordered his reserve force readied for another strike on Midway.

Meanwhile, the Kidō Butai came under attack from Midway's land-based planes, many obsolete, many flown by raw Marine recruits. To the cheers of Japanese sailors watching from below, swarming Zeros chewed the attackers up, splashing many and sending others limping back to Midway. What bombs were dropped missed their targets.

The battle's opening act had been a rousing Japanese success. The Kidō Butai was looking invincible. Its warplanes were dominating the skies. Nagumo, the victor of Pearl Harbor, seemed on his way to another triumph. Then a report came in that hit him, to quote a staff officer, like "a bolt from the blue." A search plane whose launch

IF THERE IS ONLY ONE PLANE LEFT

had been delayed that morning had sighted a U.S. task force, including a carrier, to the northeast, within striking distance.

This was indeed startling news, the most unpleasant of surprises. The Japanese hadn't expected the U.S. Navy to arrive for a couple more days. Where there was one carrier there were sure to be others. The threat couldn't be ignored. The second air strike on Midway as well as the invasion itself would have to be postponed. First, the U.S. carriers had to be sunk.

By now it was 8:30 a.m., and, except for a protective umbrella of circling Zeros, the skies over the Kidō Butai were calm again. New orders from Nagumo amped up the pace of activity aboard his carriers. On the flight decks, incoming planes returning from the Midway attack began screeching to a halt in quick succession, while below in the hangar decks, sweaty, grunting sailors carted massive armor-piercing bombs and torpedoes around as fast as their bulk would allow.

Preparing the kind of all-out strike Nagumo had in mind was going to take time. Planes returning from Midway had to be recovered and returned to the hangar decks. Much of his reserve force, which had been armed with antipersonnel bombs for the second strike on Midway, needed to be rearmed with ship-killing weapons. Then, those planes had to be moved up to the flight decks, warmed up, and launched.

To make it all happen, Nagumo needed a ninety-minute breathing spell, with no further air attacks that would trigger defensive measures and delay his preparations—just ninety uninterrupted minutes to deliver a one-hundred-plane Sunday punch that would knock hell out of the U.S. Navy.

By 9:15 he was halfway there when an officer on the *Akagi*'s bridge noticed dark specks moving low over the eastern horizon like "waterfowl flying over a lake far away." A lookout spotted them too and flashed an alarm:

Enemy torpedo planes.

No wonder John Waldron was angry. It had been an infuriating morning.

The night before the battle, unlike most of the *Hornet*'s flyers, the TBD squadron's commander had been in a buoyant mood, his eyes twinkling above a grin as he gave his pilots a final rundown on their plan of attack. A born warrior, fierce and fearless, who in an eighteen-year naval career had yet to see any action, he relished the chance to finally prove himself in combat. He told a friend he would get a Japanese carrier or he wouldn't be back.

His frustration began just after six the next morning. A search plane had sighted the Kidō Butai two hundred miles to the southwest. The hard-charging Waldron wanted the *Hornet* to launch its planes at maximum range ("I won't hesitate to run this squadron dry of gas," he'd told his men. "In that case, we'll all sit down in the water together and have a nice little picnic."), and when the launch was delayed to close the distance to the Japanese fleet, he became visibly upset.

Even worse, his TBDs would be attacking without fighter cover. The *Hornet*'s skipper, Admiral Marc Mitscher, had assigned the entire fighter escort of ten Wildcats to protect the ship's two dive-bomber squadrons. Waldron was outraged, and said so. A TBD squadron's chances were already slim enough without depriving it of fighter cover. Could his squadron get any fighter cover at all? Even one Wildcat? Mitscher's reply was a flat no.

With tensions already high, another sore subject came up: the direction the *Hornet*'s planes would be flying. Waldron's immediate superior, Commander Stanhope Ring, the *Hornet*'s air group commander, had chosen a course of 265 degrees, almost due west. Again Waldron spoke up. Flying west, he reckoned, would put the *Hornet*'s planes far north of the Kidō Butai. They would miss the enemy fleet entirely. As the discussion became heated, Mitscher cut Waldron off and told him to follow orders.

By the time the *Hornet*'s planes were aloft and heading west, with Ring's dive-bomber in the lead, it was 8:00 a.m. and Waldron was still seething. Ring was an arrogant blunderer, so hated by his pilots that several of them had discussed shooting down his plane to get rid of him. Orders or no orders, Waldron wasn't about to follow him on a wild goose chase, not with enemy carriers within striking distance. Half an hour into the flight he broke radio silence to tell Ring he was flying in the wrong direction. "I know where the damned Jap fleet is," Waldron snarled. Ring snapped back, ordering the squadron leader to stay in formation.

Breaking radio silence was a serious offense. What Waldron did next, had he survived, would have been career suicide. "The hell with you," he replied. Then he banked his TBD to the left, broke formation, and headed southwest. The rest of his squadron peeled off and formed up behind their skipper. The fifteen TBDs headed off by themselves, leaving the rest of the air group behind.

Ring doggedly continued westward for more than an hour over barren ocean, even after it became obvious he was headed in the wrong direction. Running low on fuel, his remaining pilots finally broke formation and turned back, leaving the stubborn air group commander to fly on alone, until even he was forced to swallow his pride and turn around. In what became known as "The Flight to Nowhere," none of the *Hornet*'s fighters and dive-bombers found the enemy fleet that morning. Some pilots made it back to the ship. Some ditched and were rescued. Some were never seen again. Ring himself managed to make it back to the *Hornet*. As he climbed down from his cockpit, a crowd of sailors, thinking he was returning from a bombing run, cheered him.

Meanwhile, Waldron was leading his TBDs far to the south, feeling his way by instinct. "Maybe it's the Sioux in me, but I have a hunch the Japanese ships will be in a different position than our reports have them," he'd told his pilots. "Just follow me. I'll take you to 'em."

The squadron headed toward some wisps of smoke on the horizon, and moments later a vast armada appeared, a screen of battleships, cruisers, and destroyers that seemed to stretch to the ends of the ocean. At the center of the armada were the four prizes the U.S. Navy was after, their flat, elongated silhouettes instantly recognizable, churning up mile-long wakes. Waldron had led his squadron straight to the enemy carriers.

He tried to raise the air group commander on the radio. "Stanhope from Johnny One. Enemy sighted." A pause, and again: "Stanhope from Johnny One. Answer. Enemy sighted." Ring either didn't hear the transmission or chose to ignore it. By that time the rest of the *Hornet*'s planes were too far away to help anyway. The TBDs were on their own.

Waldron's voice was calm as he addressed his men over the radio: "We will go in. We won't turn back. We will attack. Good luck." The squadron pressed on alone. No one turned back.

Ahead lay a gauntlet of forty-odd Zeros and a ring of screening ships bristling with antiaircraft guns. The TBDs were still nine miles from the nearest carrier when the Zeros jumped them. Working in tandem, the fast, agile fighters scissored the TBDs, blasting them from both sides with machine guns and twenty-millimeter cannons mounted in their wings, then circling back to blast them again. Then, as if on cue, the Zeros moved aside and the antiaircraft guns

IF THERE IS ONLY ONE PLANE LEFT

opened up. Puffs of black smoke from bursting shells peppered the sky, joined moments later by a storm of tracer fire converging on the squadron.

The slow-moving TBDs had no chance. One after another, amid a hail of gunfire, they blew up in midair or caught fire and careened into the ocean. Nagumo's chief of staff, Rear Admiral Ryūnosuke Kusaka, watched as a stricken TBD, its torpedo still on its belly, veered toward the *Akagi*'s bridge, barely missing it before crashing into the water. In the heat of battle, Kusaka paused to say a quick prayer for his gallant foe. Waldron was last seen stepping onto the wing of his burning plane as it hit the water, killing him and his gunner, Chief Radioman Horace Dobbs. Dobbs had been ordered back to the states as an instructor but had delayed the transfer to be with the squadron.

Pilots of the USS *Hornet*'s TBD squadron. Lieutenant Commander John Waldron stands third from left. Ensign George Gay is kneeling fourth from the left.

No hits were scored on the Japanese ships. Most of the TBDs never got close enough to drop their torpedoes. John Waldron would never see his beloved wife Adelaide and their two daughters again. Rusty Kenyon would never hold his baby girl. Popular, towheaded Ensign Whitey Moore, "one of the best jitterbuggers in the Navy," would never build his dream home in the West Virginia hills. Frank Polston. Otway Creasy. Grant Teats. The entire squadron, every pilot and gunner, had been killed, buried without a trace beneath the sea.

All but one.

Flying "tail end Charlie," the last plane in the squadron, Ensign George Gay had just witnessed his comrades being slaughtered. He heard his gunner, Robert Huntington, gasp, "They got me," and turned around to see him slumped over, the victim of a fatal chest wound. The young Texan who feared he would chicken out under fire suddenly found himself flying alone toward the carrier *Sōryū*, with all hell breaking loose around him and his dead skipper's words ringing in his ears—*If there is only one plane left, I want that man to go in and get a hit.*

Machine gun fire from a Zero on Gay's tail rattled against the armor plate behind his seat. Rising above shoulder level, the bullets tore past his head into his windshield and instrument panel. He squeezed a spent slug out of his arm and, with the mad thought of making it a souvenir, stuck it in his mouth. It tasted bloody.

His plane riddled with bullets, yellow tracers streaking past his wings, he tried to launch his torpedo, but the release button didn't work. The controls had been shot away. His left hand had been hit and was all but useless, so he held the joystick with his knees, reached across with his right hand, and pulled the manual release. Gay thought he heard the torpedo explode against the carrier's hull. In fact, it just missed, passing harmlessly beyond the bow.

He thought, "I've gotten rid of the goddamn torpedo. Now what do I do?" He banked his plane and flew the length of the carrier,

IF THERE IS ONLY ONE PLANE LEFT

nearly brushing the flight deck, literally eye-to-eye with a pom-pom gunner who was shooting at him. The thought flashed in his mind to crash his plane into the carrier and take it down with him, but at the last instant he pulled up. He was trying to fly free when a hit from a 20 mm shell set his engine ablaze. There was nothing to do now but ditch.

Catching a wing on the water, the TBD cartwheeled into the sea. The next thing he knew he was trapped inside five tons of sinking metal. The cockpit hood had jammed. Try as he might, he couldn't budge it. For the first time he began to panic. Seawater was rushing in around him. All the way up to his armpits. To his chin. He propped himself up on his instrument panel and with a final desperate shove forced open the hood and escaped. His leg was scorched and his hand was injured, but otherwise he was in miraculously good shape. He tried to save Huntington, but it was too late. The unconscious gunner sank with the plane. Passing Zeros were strafing Gay. He grabbed a floating seat cushion and hid his head under it. The Zeros gave up trying to shoot him and flew away.

He was lucky to be alive, but not out of danger. "Looking around, I saw the whole Jap navy steaming right down on me," he said later. "I thought they were going to run right over me."

Waldron's attack had delayed Nagumo's airstrike against the U.S. fleet eighteen precious minutes, and just as the attack was ending, as Gay's plane was hitting the water, Japanese lookouts sighted fourteen more American torpedo planes, this time approaching from the south. The airstrike would have to wait a little longer.

To the amazement—and scorn—of Nagumo and his staff, peering through their binoculars from the bridge of the *Akagi*, the planes were once again attacking singlehandedly, a second TBD squadron

Pilots of the USS *Enterprise's* TBD squadron. Lieutenant Commander Gene Lindsey stands fifth from the left. Lieutenant Tom Eversole stands seventh from the left.

being served up to the Zeros and antiaircraft gunners on a silver platter. How inept could the U.S. Navy be?

This time it was Gene Lindsey's squadron from the *Enterprise*. Through a chain of mishaps that morning—delays, faulty communication, plain bad luck—the TBDs had become separated from the carrier's fighters and dive bombers, and had arrived alone in the sky over the Kidō Butai. Lindsey led his squadron on a wide arc around the enemy fleet, stalling for time while he waited for the rest of the planes to show up. Finally, too low on gas to wait any longer, he gave the order to attack.

Just as before, while shipboard lookouts kept score, gleefully calling out each kill, the plodding TBDs were shot out of the sky. What torpedoes were dropped, the carrier *Kaga* expertly avoided. Only the fact that the Zeros were running low on 20 mm ammo prevented an outright massacre. As it was, eighteen of Lindsey's twenty-eight pilots and gunners perished. A few TBDs managed to make it back to the *Enterprise*, one so full of holes the deck crew gave up on repairing it and shoved it overboard.

"His one ambition is to fly, and already he is sprouting wings," the Annapolis yearbook had said of Midshipman Tom Eversole. "May his flight through life be a smooth one." Now he was gone, along with his gunner, the opera-loving John Lane, their premonitions of doom tragically fulfilled.

Lindsey shouldn't even have been flying that morning. Still banged up from a crash landing a few days earlier, he'd had to be helped into his cockpit. He had shown up at breakfast bruised and battered, with his ribs bandaged, moving stiffly, weak from loss of blood, looking like he belonged in the ER rather than at the controls of a warplane. "Are you really going to fly?" a messmate asked. At the Naval Academy, Lindsey had been a notorious shirker, deemed unfit to be an officer. Now, fifteen years later, he would grit his teeth and do his duty. "This is the real thing today, the thing we've trained for," he replied. "And I will take my squadron in." Piloting the lead plane, he would be among the first to die.

Following Lindsey's attack there was another break in the action, a godsend for Nagumo. At last, while his combat air patrol—his iron shield of Zeros—circled reassuringly overhead, ready to pounce if more American planes appeared, his carriers began turning into the wind to launch their airstrike. Japanese confidence, already high, was soaring. Nagumo's right-hand man, the brilliant tactician Commander Minoru Genda, was plotting the endgame in his head. The Japanese would decimate the U.S. fleet by nightfall, then subdue Midway with a night attack. Victory was just a matter of time.

It was 10:15 a.m. The Battle of Midway seemed as good as won.

GLORY, GRIT AND GREATNESS

The gunner's seat of a torpedo plane, en route to a shootout with Japanese Zeros, was hardly the ideal place to spend your twenty-first birthday. How much nicer to have been back at the *Yorktown*, toasting your entry into manhood with a cold one, preferably a root beer float from the ship's soda fountain. At the moment, though, Lloyd Childers had more important things on his mind than celebrating. Like the rest of the *Yorktown's* air group, he was busy scanning the horizon.

Held in reserve, the carrier's three-dozen TBDs, Wildcat fighters, and Dauntless dive bombers had been the last air group to depart that morning. Unlike their counterparts from the *Hornet* and *Enterprise*, they had managed to stay together, hoping to deliver a coordinated attack, a Sunday punch of their own, if only they could locate the enemy fleet, not always easy in the era before satellite tracking, when a pilot's main navigational aids were a compass, a small plotting board, and lady luck.

The *Yorktown's* planes had been aloft for over an hour; it was after ten o'clock, and their gas gauges were dropping by the minute. The last reported sighting of the Japanese fleet was more than four hours ago. By now it could be anywhere within a hundred miles.

The sharp-eyed kid from Oklahoma was the first to spot the distant wisps of smoke that could mean only one thing. Childers raised his pilot, Machinist's Mate Harry Corl, on his plane's intercom, prearranged signals went out, and soon the entire air group was headed toward the Japanese fleet, including the third and final torpedo-plane squadron that morning to take a crack at the Kidō Butai.

Within minutes, the entire Japanese combat air patrol swarmed down on the low-flying TBDs and the handful of Wildcat fighters weaving just above them, turning the surrounding sky, in the words

IF THERE IS ONLY ONE PLANE LEFT

of Wildcat pilot John S. "Jimmy" Thach, into "the inside of a beehive." When he saw a string of Zeros diving toward him, and two more squadrons streaking toward the torpedo planes from the side and front, he thought to himself: *None of us is getting out of here alive.* The badly outnumbered Wildcats were too busy fighting for their lives to protect the TBDs, leaving them to the mercy of the Zeros.

"Up ahead. Up ahead," Corl shouted over the intercom, as a group of Zeros came at him head on, shot past, then turned to attack from behind. Making pass after pass, two dozen of the speedy fighters raked the torpedo planes while Childers and his fellow gunners fired back as fast as their jam-prone guns could shoot, swinging them around on their turrets as the targets zipped by. As long as the TBDs stayed in tight formation, throwing up a curtain of machine gun fire, they held their own. When they split up to deliver a pincer attack on the carrier *Hiryu*, they became easy prey.

A hit from a 20 mm shell sent Corl's plane angling sharply downward. "I don't think we're going to make it," he warned Childers over the intercom. As the plane was about to hit the water, Corl had the presence of mind to drop his torpedo, getting rid of just enough weight to get the nose back up and stay airborne.

Planes were falling out of the sky faster than TBD pilot Wilhelm Esders could count them: "Any direction I was able to look, I could see five, six, seven, or more aircraft on fire, spinning down, or simply out of control and flying around crazily." Among the slain was the squadron's skipper, Lem Massey. The affable New Yorker was last seen trying to escape his burning cockpit just before his plane exploded.

Far off to the left, ten miles away, Esders glimpsed a "solid sheet of flames" that appeared to be a Japanese carrier on fire. It seemed the U.S. Navy had finally scored a hit. He had no time to wonder how it had happened. His target, the *Hiryu*, lay dead ahead.

He launched his torpedo, which missed, banked hard to avoid the antiaircraft fire, and was set upon by four Zeros as he fled. For twenty miles they pursued him, literally flying circles around his plane, shooting from every direction, while Esders, with his mortally wounded gunner, Robert Brazier, unable to return fire, flew on as fast as a lumbering TBD full of holes would allow, jinking his plane to dodge the tracers as best he could, listening to the hits tear into his fuselage.

The last Japanese pilot to depart flew up alongside the torpedo plane, wingtip-to-wingtip, and raised his hand with his arm bent, in what looked like a sort of half salute. "What he intended to indicate," Esders said later, "I will never know."

He managed to make his way back to the U.S. fleet and ditched nearby, pulling the dying Brazier into the plane's inflatable life raft. The twenty-five-year-old from Salt Lake City apologized for letting Esders down, drew a long, slow breath, and died as his comrade-in-arms prayed over him, leaving behind the mother who'd raised him and a wife of five months.

As had been happening all morning, not a single torpedo had found its mark—the TBDs' torpedoes, it would later be revealed, being as defective as the planes themselves. Once again, the attackers had paid a terrible price. Twenty of the twenty-four pilots and gunners in the *Yorktown*'s torpedo-plane squadron had been killed. One, Ensign Wesley Osmus, had hit the water and been captured and would soon be executed, and only one, Esders, had been rescued. The squadron had been decimated, just as its skipper had feared.

And there was still a crippled TBD in the air, trying to make it home with its engine, its elevator controls, its radio—and its gunner—shot up.

Bleeding from three gunshot wounds in his legs, his ankle shattered, Lloyd Childers kept firing his machine gun until it jammed, then blazed away at the pursuing Zeros with his Colt .45. Eventually

IF THERE IS ONLY ONE PLANE LEFT

Pilots of the USS *Yorktown's* TBD squadron. Lieutenant Commander Lem Massey is seated in the middle row third from left. Machinist Harry Corl is standing in the back row first on the left. Chief Aviation Pilot Wilhelm Esders is eighth from the left in the back row.

they broke off the attack, leaving him and his pilot Harry Corl a hundred miles from their ship with an engine that was leaking oil.

Corl somehow nursed the damaged plane back to the spot where he thought the *Yorktown* would be, only to find nothing but empty ocean. With his oil gauge nudging zero, he guessed where the fleet was and headed off in that direction, finally ditching the plane within sight of a U.S. destroyer.

He jumped in the water, and Childers rolled into it as gingerly as his wounds would allow. A rescue launch was motoring toward them. Corl watched as Childers took a few strokes toward it, then gave up, too weak to swim. All through the return flight he'd been slowly bleeding to death, and now he was starting to fade, going limp in his Mae West, struggling to lift his head, yet still alert enough to notice his plane sinking nose first beside him. Cursedly slow though the TBD might be, few planes were tougher, and this one had withstood everything the enemy had hit it with and kept flying.

The young gunner reached out and patted the tail as the plane slid beneath the water, thanking it for bringing them back.

"I saw this glint in the sun. It looked like a beautiful silver waterfall."

That is how Wildcat pilot Jimmy Thach, glancing upward from a dogfight, described the climax of the Battle of Midway, the moment when, after a morning of foul-ups and failed attacks, the U.S. Navy caught lightning in a bottle and the Japanese Navy, on the brink of victory, caught hell.

The *Yorktown*'s Dauntless dive bombers had slipped into the unguarded sky high above the Kidō Butai while the Zeros were off tangling with the TBDs and Wildcats down near the water, and by pure chance, after searching for the enemy for two hours, the *Enterprise*'s Dauntlesses had shown up from a different direction at precisely the same moment.

At 10:25 a.m., with no enemy fighters around to harry them, fifty silver-colored dive bombers, gleaming in the late-morning sun, pushed over into their bombing runs and came cascading down past the clouds. ("I'd never seen such superb dive bombing," Thach said. "It looked to me like almost every bomb hit.") By ten thirty, three Japanese carriers were ablaze and would soon sink, taking with them Dai Nippon's dreams of empire. Before the battle was over, the Americans would sink the fourth Japanese carrier, and the Japanese would sink the *Yorktown*, making the final tally four downed carriers to one, a stunning victory for the U.S. Navy, and a disaster from which Japan would never recover.

IF THERE IS ONLY ONE PLANE LEFT

The Battle of Midway didn't end the war in the Pacific—it would take three more bloody years before Japan surrendered—but it stopped Japanese expansion cold, giving the United States the breathing spell it needed to bring its huge industrial and manpower reserves to bear. After Midway, an American victory was just a matter of time.

Even in a conflict in which, said Admiral Chester Nimitz, uncommon valor was a common virtue, the TBD squadrons at Midway stand out. Their death rate alone, shocking as it was—all but a handful of the eighty-two pilots and gunners were killed—doesn't do their heroism justice. Knowing how hopeless the odds were, they went aloft in outmoded planes and flew into the teeth of withering fire, never wavering—*We will go in. We won't turn back. We will attack*—laying down their lives in doomed attacks, delaying Nagumo's counterstrike and drawing away the Zeros, paving the way for the dive bombers to come in and deliver the crushing blow that changed the course of the war.

"I am not a hero," George Gay would claim in his memoirs. A grateful nation disagreed. Rescued by a PBY patrol plane after spending a harrowing day in the water hiding in plain sight from the Japanese fleet, the lone survivor of the *Hornet*'s TBD squadron became a national hero, making the cover of *Life* magazine and dividing his time between combat missions over Guadalcanal and publicity appearances for the Navy. After the war, the engaging Texan flew airliners, published his memoirs, and became a sought-after speaker.

When he died in 1994 at age seventy-seven, his ashes, according to his wishes, were flown out to the middle of the Pacific and scattered over the place where he and his squadron mates had been shot down.

GLORY, GRIT AND GREATNESS

Bill Esders saw action in the Solomon Islands where he won the Distinguished Flying Cross to go with the Navy Cross he'd won at Midway. The former enlisted man retired from the Navy a full commander and passed away in 1994 at age eighty.

Harry Corl, the only other TBD pilot from the *Yorktown* to survive the battle, was shot down and killed three months later while making a bombing run on a Japanese cruiser.

Dive-bomber pilot Dusty Kleiss avenged Tom Eversole by hitting two Japanese carriers with bombs during the Battle of Midway. Kleiss passed away in 2016 at the age of one hundred, mourning his friend to the end.

A lucky break saved Lloyd Childers' life. The destroyer that rescued him may have been the only one in the fleet with a doctor aboard. With his patient laid out on a mess table, he staunched Childers' bleeding just in time.

After a long convalescence, Childers fulfilled his dream of becoming a combat aviator, and a highly decorated one at that, flying Corsairs in Korea and commanding a helicopter squadron in Vietnam for the Marine Corps, before retiring as a lieutenant colonel. As a civilian, he earned a Ph.D. and became an administrative dean at a small college in California, then retired for good after twenty years to play golf and spend more time with his family. (Childers' advice to an interviewer: "If you have a wife, treasure her.") He passed away in 2015 at age ninety-four.

On the battle's seventy-second anniversary, Childers's ninety-third birthday, a reporter asked him for his thoughts on Midway.

The hero who had fought so gallantly and nearly bled to death, the distinguished scholar who would always be a kid from Oklahoma at heart, drawled a matter-of-fact reply: "We beat 'em up pretty good, but it sure took a lot out of us."

CHAPTER 2

THE MOST HATED MAN IN AMERICA

John D. Rockefeller

May 26, 1937
Pocantico Hills, New York

Condolences poured in from governments all over the world, from Canada and Japan and Italy and Nicaragua.

The Chinese sent their "deepest sympathy" for a "friend" whose "memory will long endure." Budapest, Hungary, passed a resolution honoring him, and in Paris, France, outside the Palace of Versailles, which had been refurbished with his money, flags flew at half-staff and black crepe hung from the lampposts. Oil workers across the globe stood in silence for five minutes.

The legendary oil king was dead at age ninety-seven.

In his home country, the United States, along with tributes from those who knew him, there was an air of tension. At the private funeral, attended by family and friends at his estate in the wooded hills north of New York City, security was tight. State troopers patrolled the grounds and stood guard as mourners filed through the main gate. The next day, at the family's cemetery plot in Cleveland, Ohio, the casket was lowered into a bombproof tomb. Guards would be posted around the clock.

Branded the most ruthless robber baron of America's Gilded Age, the ogre in a skullcap who crushed his competitors and ruined people's lives, he was the most controversial man in America, and perhaps the most hated. The press demonized him. Cartoonists lampooned him. Politicians railed against him. Anarchists threatened to blow him up.

Yet the public never really knew the soft-spoken gent who preferred family bike rides to board meetings, church picnics to high society. Or the pious genius who changed the world. Surrounded by security guards, pursued by process servers, a revolver beside his bed, he wanted only to be left alone to carry out what he saw as his sacred calling: to make more money, and give more away, than anyone had ever dreamed of.

"God gave me my money," John D. Rockefeller was fond of saying, "because [He] knew that I was going to turn around and give it back."

The road to the world's largest fortune began when, to help support his family, John dropped out of high school to look for work, pounding the pavement in Cleveland's bustling business district until his feet hurt. "I did not go to any small establishments," he remembered. "I was after something big." He settled for a job as an assistant bookkeeper for fifty cents a day. Thrilled to be hired, he skipped down the street like a lovestruck schoolboy. He would celebrate the date—"Job Day" he called it—every year for the rest of his life.

The eager youth pored over the account books day and night, loving the musty smell of the ledgers and burning with ambition.

"I'm bound to be rich, bound to be rich, *bound to be rich*," he once exclaimed, slapping his listener's knee. He rose to chief bookkeeper, kept a ledger of every nickel he spent (looking back, he would chide himself for spending $2.50 on a pair of fur gloves), and banked his salary, awaiting the day he could start his own business.

When his boss refused to give him a raise, Rockefeller quit. He borrowed $1,000, chipped in the $800 he'd saved, found a partner, and started up a commodities company. Now penniless and without a steady paycheck, he would recall feeling exhilarated: "It was a great thing to be my own employer [and] I swelled with pride." He felt he had the world by the tail. He was nineteen years old.

A Midwest crop failure hit the new business hard. Rockefeller took to the road, crisscrossing Ohio and Indiana in search of customers. He proved himself a master of the soft sell. The youngster's quiet poise impressed people. Orders started rolling in. That first year the firm turned a then-tidy profit of $4,400. To a kid who'd grown up in shabby clothes sharing a bed with his brother, it was all the money in the world. Before long, it would be pocket change.

That summer, August 1859, from the wilds of northwest Pennsylvania came startling news. A former railroad conductor named Edwin Drake—thought to be insane—had drilled a well down through the bedrock, and a few barrels of crude oil had come bubbling up, triggering a wave of black-gold fever like nothing since the days of California's mother lode. Overnight, forests disappeared and derricks and boomtowns sprang up.

What Rockefeller would call one of God's "bountiful gifts" at times seemed more like a curse. Oil wells had a nasty habit of exploding. Gushers turned to dusters in the blink of an eye. For every wildcatter who struck it rich, dozens went bust. On potholed trails knee-deep in mud and oil, barrel-laden wagons foundered, and exhausted dray horses collapsed and drowned in the muck.

But once the oil made it out of the wilderness, it could be refined into something that, in the days before light bulbs, was indeed a godsend: clean, bright-burning kerosene. Rockefeller smelled money in oil—everyone did—but the sober-minded businessman wanted no part of the crapshoot that was wildcatting. So he and two partners opened a small refinery in the farm country outside Cleveland alongside grazing cows.

"I shall never forget how hungry I was in those days," he later wrote. "I stayed out of doors day and night; I ran up and down the tops of freight cars. . . . I hurried up the boys."

With the American Civil War underway, prices soared. Rockefeller, like many young men of means, hired a substitute to join the army in his place, and sat out the war at home, dealing in kerosene and foodstuffs, and raking in profits. To support the Union cause and ease his conscience for not enlisting, he paid to outfit a small company of soldiers. By the war's end, he was on his way to being rich.

After the war, as the volatile oil market drove other refiners out of business, Rockefeller was cutting waste and prospering. Rather than ship kerosene in expensive, ready-made barrels, he bought stands of oak trees, hired coopers, and built his own, with the staves shortened to save a few extra pennies. He bought oil directly from the drillers, bypassing wholesalers. While other refiners were throwing away their oil byproducts, he and his partners used theirs to make paint, varnish, vaseline, candles, and scores of other items. Even chewing gum. He even decreed the number of drops of solder his welders could use to seal a five-gallon kerosene can—exactly thirty-nine—then had the drops that fell on the floor collected and reused.

He was just as meticulous in his business dealings. "If there was a cent due us, he wanted it," a partner said. "If there was a cent due the customer, he wanted the customer to have it." Rockefeller became known as someone who lived up to his promises and paid his debts.

By the time he reached his early thirties, his Standard Oil Company of Ohio was the country's largest kerosene producer, and he was among Cleveland's leading businessmen, with a spacious brick house on the city's millionaire's row.

It was a far cry from his childhood in rural New York, where he'd been born John Davison Rockefeller in 1839 of what he called good old American stock: German, English, and French on his father's side, Scotch-Irish on his mother's. The boy was so quiet that years later former neighbors would have trouble remembering him. He was no better than an average student, described by a family tutor as "just an ordinary well-behaved boy plodding along with his lessons."

Yet the lad had a knack for making money. By age seven, while other children were busy learning to read and write, he was catching wild turkey chicks and raising them for sale, and buying candy by

Top left: John D. Rockefeller age 18. *Top right:* Rockefeller in his mid-forties.
(COURTESY OF ROCKEFELLER ARCHIVE CENTER.)

Pennsylvania oil fields, circa 1865. (LIBRARY OF CONGRESS.)

the pound and selling it by the piece to his siblings at a healthy markup. At Sunday services, at the urging of his mother, he would drop one or two of his hard-earned coins into the collection plate, the beginning of what would become hundreds of millions of dollars in charity.

His traveling salesman father was on the road much of the time, leaving John's mother to raise their six children and eke out a living on the family's small farm. John never forgot shivering in the cold as he rose before dawn to milk the family cow. Even as a multimillionaire, the former farm boy wouldn't mind getting his hands dirty and would often be seen at one of his factories early in the morning, coat shed, sleeves rolled up, pitching in beside his workers.

When William Rockefeller wasn't away peddling his wares, he owned and operated a logging company and was a horse trader and money lender. "Devil Bill" wasn't much of a role model—he was a

John D. Rockefeller's parents Eliza Davison Rockefeller and William Rockefeller. (COURTESY OF ROCKEFELLER ARCHIVE CENTER.)

scoundrel and a bigamist, with a second family in Canada—but he had a good head for business, a skill he taught John and his brothers by bilking them. "I trade with the boys and skin 'em," he bragged. "I want to make 'em sharp."

John's mom believed in Christian charity and motherly affection. And whippings. "I'm doing this [out of] love," Eliza Rockefeller would say as she was tanning his backside with a switch. He would inherit from her self-discipline, a cool head, and an oft-remembered adage: Willful waste makes woeful want.

Ignoring experts who predicted the oil fields would soon run dry, John D. recapitalized Standard Oil and started buying up refineries, dozens of them, first in Cleveland, then in Pittsburgh, New York, and other cities. With kerosene prices slumping, his competitors had little choice but to sell. Some would accuse him of forcing them to sell for less than their refineries were worth. In fact, considering how badly the industry was struggling, he paid a fair price, often more. Refiners who took payment in Standard Oil stock rather than cash, and held on to it, made a killing.

By the late 1870s, Standard Oil controlled nearly 90 percent of America's kerosene production, and Rockefeller had turned a chaotic industry into a well-run conglomerate. Employees were cross-trained, promotions merit-based. Former owners were kept on to manage their refineries, and in some cases they became Standard Oil executives. Workers received generous wages and pensions, stock options, and bonuses, sparing Standard Oil the labor trouble of other companies. Even one of Rockefeller's harshest critics admitted the company had a "remarkably loyal and interested work force."

There were also backroom deals. When word leaked that the railroads were giving Standard Oil hefty rebates on shipping fees, other refiners cried foul. Shipping rebates would become Rockefeller's most notorious scandal, a never-ending supply of red meat for his critics. Yet, such rebates were common in the oil industry. Some of the same refiners who were crying foul were getting them. Standard Oil received the biggest because it guaranteed minimum shipments, loaded and unloaded its own barrels, and, most importantly, was by far the railroads' biggest customer. A railroad executive put it simply: If you ship as much oil as Standard, you'll get the same rebates.

It was the promise of an even better railroad deal that drew Rockefeller into the infamous cartel of railroads and refiners called the South Improvement Company. The scheme to manipulate the rates for shipping oil, raising them for every customer and then kicking money back to the cartel's members, though legal at the time, spread alarm across the Pennsylvania oil fields. Thousands of protestors took to the streets. Standard Oil barrels were vandalized with a skull and crossbones. A local newspaper branded Rockefeller "the Mephistopheles of Cleveland." Ironically, the cartel collapsed before it really got started, the victim of an oil producers' boycott and the cancellation of its charter by the Pennsylvania legislature. Rockefeller came away with nothing but a tarnished reputation.

In person, he was nothing like the haughty plutocrat he was thought to be. "The bigger he grew," one of his buyers said, "the more modest he became." He would lie awake at night and warn himself not to become too self-important. (Ar*e you going to let this money puff you up?*) He was punctual because he thought it was rude to make people wait, and he treated his workers with respect, encouraging them to submit their grievances to him, complimenting them on their work, and greeting them, according to one worker, with "a nod and a kind word."

The shy, quiet youngster became a man who prided himself on listening more than he talked. At board meetings he would surrender the head chair and say little. Sometimes he would step away from the table and doze on a couch. Sometimes the only thing he said during a meeting was "Goodbye," as he was leaving.

His self-control, inherited from his mother, was legendary. Through seventy years of Wall Street panics, bad press, death threats, and antimonopoly crusades, he never lost his temper, never swore (the closest he came was an occasional "Pshaw!"), never got flustered, and never drank a drop. During high-level negotiations, he wore an unshakable poker face, never letting anyone see him sweat.

Looking back, he would admit that stress had taken its toll. For years he didn't have a good night's sleep. In troubled times he turned to his most trusted advisor, the love of his life.

The young suitor had spent nearly a decade, ever since high school, courting erudite, well-born Laura Celestia "Cettie" Spelman, and after seven years of marriage was still writing her poignant letters: "I dreamed last night of the girl Celestia Spelman and awoke to realize she was my Laura."

When she finally married him—he was twenty-five; she was twenty-four—she gave up a career teaching school to devote herself to raising a family and to her husband's career. She had a keen business mind (keener than his own, Rockefeller said), and he often sought her advice on business matters. Early in their marriage, before she became an invalid, he would bring home the company books for her to peruse. She was also his soulmate, a sympathetic ear for an intensely private man loath to reveal his vulnerable side to anyone else. "I feel like a caged lion and would roar if it would do any good," Rockefeller wrote her from New York during the South Improvement ordeal. "How much I would give for wings to reach you tonight.... The world is full of Sham, Flattery and Deception; the home is a haven of rest and freedom."

Rockefeller's haven in those days was Forest Hill, his estate a few miles and a world away from the Standard Oil offices in downtown Cleveland. John and Cettie's home there was an ornate Victorian manor ("a monument of cheap ugliness," journalist Ida Tarbell called it), but what enchanted him were the surroundings: the panoramic view of Lake Erie and the wooded acreage crisscrossed by streams and trails, a setting that reminded him of his boyhood home in New York's lake district. He delighted in landscaping the grounds, putting in foot bridges and gravel pathways, and transplanting trees to open up the glorious view.

Often he would leave work early and gallop back to the estate in his horse-drawn buggy to be with Cettie and their four children. There was a pond where he taught the kids to swim and ice skate, and when the moon was full he led them on night bike rides along the winding trails. Indoors, he played blindman's bluff with them

Top left: Laura Celestia ("Cettie") Rockefeller, née Spelman. *Bottom*: Forest Hill, the Rockefellers' mansion outside Cleveland. (BOTH IMAGES COURTESY OF ROCKEFELLER ARCHIVE CENTER.) *Top right*: Journalist Ida Tarbell.

and, to their delight, balanced things on his nose, things like plates from Cettie's best set of China, or crackers that he flipped into the air and caught in his mouth.

Family time went hand in hand with religion. Each day began with family Bible readings before breakfast—late arrivers were fined a penny—and ended with Cettie watching over the children's bedtime prayers. The Rockefellers' social life revolved around their church rather than the theater, gala balls, and exclusive clubs normally enjoyed by the ultra-rich. It was the same middle-class Baptist church John had been attending since he was a teenager, where he was now lead trustee and head of the Sunday school.

He liked the Bible-thumping Baptist sermons ("I need good preaching to wind me up, like an old clock.") and felt more at home among the congregation of ordinary working folk than he would have among the well-heeled faithful of the Episcopal church up the street. After belting out hymns in his baritone voice at Sunday services, the oil baron stayed to mingle with the tradesmen and shopkeepers he'd prayed with, quietly pressing envelopes of cash into the palms of the needy as he shook their hands, ignoring the interlopers who showed up to gawk at him.

When a reporter buttonholed him and quoted Luke 18:25—"For it is easier for a camel to go through a needle's eye than for a rich man to enter into the kingdom of God"—Rockefeller fired back with Proverbs 22:29—"Seest thou a man diligent in his business? He shall stand before kings."

Now middle-aged and rich beyond his dreams, the go-getter who had worked day and night adopted a new creed: "Go along at an even pace, and keep from attempting too much." Working less while accomplishing more, he revolutionized the oil industry. He pioneered the use of storage tanks and railroad tank cars—scrapping the unwieldy wooden barrels—and built a sprawling network of pipelines.

One idea seemed sheer folly. When he proposed buying vast tracts of a new oil field near Lima in western Ohio, his partners "held up their hands in holy horror." The sulfurous Lima oil fouled machinery and gave off a nauseating stench. No one wanted to touch it, let alone buy it. Rockefeller stowed away forty million barrels of the "skunk oil" and bided his time. Two years later, when a chemist discovered how to cleanse it, Rockefeller—and his partners—made a fortune. One of them marveled at his foresight: "In business we all try to look ahead as far as possible, but Rockefeller always sees a little further ahead than any of us, and then he sees around the corner."

Peering around the corner, he set his sights overseas. Despite stiff competition in the Orient from the companies that would later merge into Shell Oil, and from the vast Russian oil field on the Caspian Sea, Standard Oil captured two-thirds of the world's oil trade, shipping its kerosene not only throughout Europe but to the four corners of the globe—to the Middle East on the backs of camels, in oxcarts to India's northern reaches, and up China's Yellow River in sampans.

By the early 1880s, Standard Oil was the world's largest business enterprise, dominating the oil industry with a hundred-thousand-man workforce, thousands of derricks and tank cars, a fleet of tankers, and pipelines that spanned the eastern United States, feeding giant refineries built to quench the world's thirst for oil.

To keep the operation running, Rockefeller needed a way to skirt laws that restricted a corporation's ownership of out-of-state companies. The solution was to convert Standard Oil and its dozens of subsidiaries into a giant "trust." Shares of stock were redeemed for trust shares. Corporate directors became trustees.

Although good for profits—Rockefeller's fortune would quintuple in the next dozen years—the move was a public-relations disaster. He'd created a behemoth reformers were determined to slay. Trust-busting politicians from both sides of the aisle campaigned on promises to break up Standard Oil.

"Fighting Bob" La Follette, Wisconsin's liberal Republican governor, called Rockefeller "the greatest criminal of the age." Democratic presidential candidate Williams Jennings Bryan said Rockefeller belonged in prison. States passed antitrust laws and went after Standard Oil in court. Dodging subpoenas, the oil king went on the lam, pursued by process servers and reporters in a headline-making manhunt.

Political cartoons depicted Standard Oil as a hog or a giant octopus or a toga-clad socialite reclining on the backs of the common people, and Rockefeller as an anaconda or Goliath wielding a club.

Muckraking journalists such as Henry Demarest Lloyd and Ida Tarbell savaged him in print. Tarbell called him a "pitiful monstrosity" with "unclean flesh," a "monk of the Inquisition" who forced his rivals "into a pit of unknown horrors." Laced with half-truths and outright errors—among other things, she repeated the myth that Rockefeller had fleeced a widow with three children out of their refinery—Tarbell's series on Standard Oil for *McClure's* magazine, later published in book form as *The History of the Standard Oil Company*, raised the outrage to a fever pitch. Threats of kidnapping and bombing became part of Rockefeller's daily life, prompting him to beef up the security around his household and sleep with a revolver beside his bed.

"Run or Rockefeller will get you," mothers playfully warned their toddlers. Behind the bogeyman image was a pragmatic businessman, as inclined to negotiate as he was to play hardball. He once bought connecting strips of land that ran all the way across

Rockefeller and Standard Oil were prime targets for political cartoons.

Pennsylvania to block the path of a rival pipeline. When it was finished, snaking its way along a ravine he'd overlooked, he struck a deal with the owners to divvy up the oil traffic.

He lobbied against laws harmful to Standard Oil, waged price wars, and gathered data on the sly about his competitors' shipments. Meanwhile, his rivals were lobbying for government action against Standard Oil, and forming cartels and waging price wars of their own. In Rockefeller's shark-infested world, hard-nosed tactics were a fact of life. Oil was a cutthroat business.

As he prevailed in the battle for the oil industry, kerosene prices plummeted. When he founded Standard Oil in 1870, the price was twenty-five cents a gallon. Twenty-four-years later, partly due to new oil fields opening up, but also due to his cost cutting, he was churning out over a billion gallons of high-quality kerosene a year for less than a penny a gallon, and selling it for a nickel.

Good lighting was no longer a luxury reserved for the rich. All over the world, working-class families who had lived at night by the dim firelight of a hearth lit up their homes for a penny an hour with what Rockefeller called his "poor man's light."

As electric lights replaced kerosene and cars appeared, Standard Oil shifted to making gasoline, opening America's first filling station and fueling the Wright brothers' maiden flight at Kitty Hawk. Trains and tractors chugged along on Standard's gear grease and axle grease. Henry Ford's original horseless carriage tooled around on Standard's Atlantic Red motor oil. If steel magnate Andrew Carnegie can be said to have built America, Rockefeller was the one who made it run.

He used his immense wealth to cure disease: meningitis in New York City, hookworm in the deep South, yellow fever in Central America, and malaria in Italy. He bankrolled the University of

Chicago and Colonial Williamsburg, fought malnutrition in New Guinea, and built the League of Nations library in Geneva and a hospital in Peking. One of his pet projects was Spelman College in Atlanta, a school for African American women where Martin Luther King Jr.'s mother and grandmother studied. But he was proudest of his Rockefeller Institute for Medical Research, responsible for one Nobel Prize—for work that paved the way for organ transplants—and deserving, according to the Hearst press, of half a dozen others. Over his lifetime, Rockefeller's charitable gifts totaled $550 million, a mind-boggling sum in those days.

Yet no matter how much good he did, people still hated him. They refused to believe someone supposedly so ruthless could be so caring. A myth arose that his crusade against hookworm, a parasite spread by walking barefoot through contaminated soil, was part of a scheme to start up a shoe business in the South.

His image wasn't helped by the fact that, to put it mildly, he didn't age well. By the time he turned sixty, a rare skin disease had made all his hair fall out. Gone were his full head of auburn hair, his neatly trimmed walrus mustache, even his eyebrows and eyelashes. In cartoons, the skullcap he wore to hide his baldness became a tiny pillbox hat perched atop a shiny, bulbous pate. The flat gray wigs he carpeted his head with didn't look much better.

Tall for those days at five-foot-eleven, trim and broad shouldered, robust looking well into middle age, by his seventies, possibly due to a Spartan diet necessitated by stomach trouble, he became gaunt and shriveled. He would remain an avid golfer and lover of the outdoors into his nineties, but his crepe-paper skin stretched over a hairless, bony face made him look like a mummy.

Whatever his age, whether he was checking a factory floor for stray drops of solder, or balancing something on his nose, his blue-gray eyes never stopped studying everyone and everything. His nickname in the oil business was The Sponge. Seeing his eyes in pictures, you can almost hear the wheels spinning in his head.

The day he finally answered a subpoena and appeared in Chicago to testify, a phalanx of cops held back the unruly crowd. As he walked past, people reached out and ripped buttons off his blazer. Unfazed—"the coolest looking man" in the courtroom on that sweltering July day, a reporter wrote—he was calm and polite during his brief testimony, and afterward bantered with the press. What did Rockefeller think of Chicago? *That it will become the greatest city in the world.* Did he have any advice for people? *The way to ensure happiness is to be good.*

The *Chicago Tribune* gushed with praise: Rockefeller was "deferential to the court . . . friendly with the curious public that crowded

Left: Sixty-nine-year-old Rockefeller in 1909, his hair claimed by a skin condition. (LIBRARY OF CONGRESS.) *Right*: Rockefeller in his early eighties.

his path . . . affable with reporters, genial to all comers, and at peace with the world."

For thirty years, busy with his family and career, secretive to boot, he had shied away from the spotlight, turning away reporters at his doorstep, refusing, against Cettie's advice, to answer his critics. "Let the world wag," he'd snorted. Now, fed up at last with bad publicity, he emerged from his shell. He published his ghostwritten memoirs and started golfing with reporters.

The results were mixed. One paper accused him of acting like a politician. Some reporters were surprised how likable he was. "It is barely possible that the curious old man has been misrepresented," one wrote, "and that the world owes him an apology."

The president of the United States disagreed. Teddy Roosevelt, whose trust busting would earn him a place on Mount Rushmore, went after Standard Oil like he was charging San Juan Hill again at the head of the Rough Riders.

The result was the mother of all antitrust cases. Finally, in 1911, after four hundred witnesses and seemingly endless appeals, thirty years after Rockefeller had formed the Standard Oil trust, the U.S. Supreme Court ordered it dissolved. "Dearly beloved, we must obey the Supreme Court," he wired his colleagues. "Our splendid happy family must scatter."

But John D. would get the last laugh on T. R. On learning of the ruling, Rockefeller turned to his golfing partner that day—a priest—and said, "Father Lennon, buy Standard Oil." Had the father not taken a vow of poverty, the tip might have made him rich. The trust, it turned out, was worth far less than the sum of its parts. Rockefeller's holdings in spinoffs created by the breakup, including companies like Mobil Oil, the Continental Oil Company (Conoco), and Standard Oil of New Jersey, later Exxon, sent his net worth soaring.

Roosevelt groused that all he'd done was double Rockefeller's already huge fortune. In fact, it probably tripled. No one knows exactly how much Rockefeller was worth, but losing the antitrust case may have made him America's first billionaire. Blowing by Andrew Carnegie, he became the richest man in the world.

Burned out on the oil business, he retired and became a golf fanatic. He built himself two golf courses, bought a third for good measure, and played five hours a day, rain or shine. "Yesterday morning I played with the thermometer at twenty," he proudly wrote his niece. "It was cold indeed . . . but a good thing for my health." To perfect his swing, he anchored his feet under croquet wickets and hired a boy to stand next to him and bark out, "Keep your head down. Keep your head down." After a solid *THWACK*, Rockefeller would exclaim, "See! See! Method. Method." A slice would leave him shaking his head and muttering, "Shame. Shame." Through sheer persistence, he whittled his handicap down to four.

The aging teetotaler behaved at times like a happy drunk, dancing a jig when he won at shuffleboard, and telling—and retelling—corny jokes, like the one about the lunatic in the asylum who told a visitor, "If you can't get me out of here, bring me a piece of dry toast. I'm a poached egg." Once at a dinner party, to prove a point about corns, Rockefeller stripped off a shoe and sock and plopped his bare foot onto the dinner table for all to see.

"Bless you. Bless you. Bless you." he would say, bowing repeatedly as he handed out shiny new dimes to onlookers. The practice probably began as a way to break the ice with his neighbors in Pocantico (puh-KAN-ti-co) Village outside his New York estate. Eventually, a new generation of Americans would know him from newsreels and publicity photos as the geezer in a gray toupee who gave a

Rockefeller hands a nickel to a two-year-old.

"Rockefeller Dime" to strangers, golf partners, fellow celebrities, whomever he encountered. Children got a nickel and sometimes an extra penny, along with some grandfatherly advice: "The penny is to spend; the nickel is for your bank."

Busy enjoying life, he left Cettie's side more than he should have. While she lay bedridden from a long list of ailments—asthma, emphysema, colitis, sciatica, anemia, and what was probably a stroke—he took off for months at a time on golfing trips, leaving her to the care of family and nurses.

He was off vacationing with their son and daughter-in-law when word came that Cettie had died quietly at home. She and her husband had long since become polar opposites. As John had grown more lighthearted with age, his once vivacious bride had become a stern Christian fundamentalist. Still, fifty years of marriage had formed an indelible bond. The couple never lost their affection for each other. When John was home, he'd slip away from his dinner guests to Cettie's bedside to bring her a flower from their garden. Among her most prized possessions was the three-dollar ring he'd bought her when they were married.

The day she died, their son witnessed something never seen before. The former titan of industry renowned for his self-control broke down and wept.

He'd lost his soulmate but not his zest for life. For a change of scenery, he moved to the coast and spent his last years as the most eligible bachelor in Ormond Beach, Florida. He liked the ladies and they liked him. Even at ninety, the world's richest man had no trouble getting dates.

Women lined up for a chance to ride along on his daily drives in the country. Rockefeller would squeeze into the backseat of his town car between two buxom spinsters, with a big blanket, a "lap robe," draped over the threesome up to their necks. What took place under the blanket, as the car wound its way through the countryside and the chauffeur kept his eyes glued to the road, is anyone's guess. On one of the drives, when the motorcade came to a stop sign, one of Rockefeller's dates jumped out of his car and scurried to the lead car shouting, "That old rooster. He ought to be handcuffed." Even after that story made the rounds, what was known as the "hot seat" was as much in demand as ever.

When not canoodling, he was the venerable gent-about-town, strolling the waterfront in his linen suit and boater, cane in hand, chatting with the townsfolk and passing out dimes, being greeted by a quaint nickname he'd thought up for himself. In Ormond Beach he would be forever known not as the oil king or the world's richest man, but as Neighbor John.

His secret to longevity? A spoonful of olive oil a day. Or was it plenty of fresh air and sunshine? Or plenty of rest? Or a calm personality? He gave a different answer every time the subject came up. Good genes were the real secret. His father had lived to be ninety-five. One of the oil baron's grandsons, David Rockefeller, would live to be a hundred and one.

John D. Rockefeller had been born in the days of mule teams and sailing ships, when a fifty-mile stagecoach ride took eight hours. By the time he died at ninety-seven, race cars were setting speed records on the Daytona flats a few miles from his home, and transatlantic flight was becoming common. Few people had lived to see so many changes. No one had done more to bring them about.

His bland last words, "Raise me up a little bit," murmured to his valet, were typical of him. In death, as in life, he kept his innermost thoughts to himself. It would be left to a member of the African American church Rockefeller attended in Ormond Beach to provide his epitaph. "He was a fine, gentle soul and a real Christian," the man told the *New York Times*. Then he pointed to an announcement that the twenty-third Psalm would be read at the Sunday service in Rockefeller's honor.

"It was his favorite. 'Even though I walk in the valley of the shadow of death I shall fear no evil.'"

CHAPTER 3

RAYS OF GLORY*

The American Army at Trenton

In December 1776, just five months after the thirteen American colonies had declared their independence from Great Britain, the American Revolution seemed doomed.

Former rebels, including some members of the Continental Congress, were lining up in droves to swear allegiance to King George III. The new nation's paper money was all but worthless.

And on the battlefield, the American army was near collapse.

Driven out of New York, across New Jersey, and all the way into Pennsylvania, decimated by battle losses, disease, and desertion, outmaneuvered and outfought by General William Howe and his crack British redcoats and Hessian mercenaries, the rebels were down to a few thousand freezing, half-starved troops. General Howe

---

* "You will think me transported with enthusiasm, but I am not. I am well aware of the toil and blood and treasure that it will cost us to maintain this Declaration and support and defend these states. Yet, through all the gloom, I can see the rays of ravishing light and glory." —John Adams

had only to brush them aside and march into the rebel capital, Philadelphia, and the Revolution would be over, the dream of independence dead in its cradle.

But when Howe hesitated, the American commander, General George Washington, decided to attack. On Christmas night in a blizzard, he and his men would slip across the ice-choked Delaware River into enemy territory.

Ready to go down fighting in a sleepy little town called Trenton.

Back in late June, New Yorkers had watched in grim silence as British masts appeared off Long Island, so many of them that, to one rebel soldier, lower New York Bay resembled a pine forest. "I could not believe my eyes," Private Daniel McCurtin wrote in his diary. "I thought all of London was afloat."

The rebels had awakened a sleeping giant. After being humiliated in some small, curtain-raising battles with legendary names—Lexington, Concord, Bunker Hill, Fort Ticonderoga—the British were looking to crush the Revolution once and for all. They had sent one of the deadliest armies ever for its size, a thirty-thousand-strong force built around elite units: grenadiers, the army's shock troops, masters of the bayonet; mobile light infantry, skilled in wilderness fighting; and the Royal Highlanders, a regiment of fierce, barrel-chested Scotts whose weapons included throat-slitting daggers tucked inside their stockings. Last but not least was a contingent of several thousand highly trained and disciplined Hessian mercenaries, well worth the price tag of £3,000,000 paid to the ruler of the German province of Hesse-Cassel.

Up against these razor-sharp professionals was a force of comparable size but woefully undertrained, a ragtag array of citizen soldiers: farmers, shopkeepers, merchants, tradesmen, fisherman, hillbillies, and college students. The American army, sniffed one British general, was nothing but "a rabble in arms."

Its leader was a Virginia planter, a former militia colonel who had never before commanded an army. "I do not think myself equal to the command I am honored with," George Washington had confessed to Congress.

Jubilation over the signing of the Declaration of Independence on July 4th gave way to a disastrous late summer and fall. At the Battle of Long Island in New York, General Howe, a master tactician, smashed Washington's flank, sending the terrified rebels fleeing all the way to Brooklyn Heights. When the British landed at Kip's Bay in Manhattan, the rebels threw down their muskets and ran. At Fort Washington on the Hudson River, the three-thousand-man garrison was overrun. Watching from across the river as his surrendering men were being bayoneted, General Washington lowered his spyglass and wept.

Washington and Staff at Fort Lee, Watching the Battle of Fort Washington by John Ward Dunsmore.
(COURTESY OF FRAUNCES TAVERN MUSEUM, NEW YORK CITY.)

GLORY, GRIT AND GREATNESS

By late November his army was in full retreat, fleeing across New Jersey, said one redcoat, "like scared rabbits." Forced to abandon their supplies, their rations reduced to a few handfuls of raw flour a day, the rebels began to starve. Months of campaigning had reduced their clothes to rags. Men without shoes hobbled on feet wrapped in cloth or strips of rawhide.

Before long, typhoid, dysentery, scurvy, typhus, smallpox, and pneumonia were rampant. Stricken men lucky enough to survive were too sick to fight. The Delaware Continentals, one of Washington's best regiments, was down from seven-hundred-fifty men to ninety-two.

Some men quit when their yearlong enlistments were up. Others deserted. So many were leaving that the roads became clogged with ex-soldiers heading home. By mid-December, Washington was down to six thousand effectives and was confiding to his brother, "If every effort is not strained to recruit a new army with all possible expedition, I think the game is pretty near up."

As the rebels crossed the Delaware River into Pennsylvania and fell back toward Philadelphia, panicked citizens piled their furniture onto wagons and fled. "Fellow citizens our enemies are advancing upon us," city fathers proclaimed. "We [must] unite in this time of extreme danger." Echoing the rallying cry, Congress voted to remain, then promptly fled to Baltimore.

Throughout the long retreat, the British had nipped at the rebels' heels but had come no closer. Howe's second-in-command, General Henry Clinton, urged him to march the British troops on Staten Island northwest to head Washington off.

Howe ignored the advice. He had a dozen reasons for not chasing the rebels down and crushing them. The campaign had worn out his army. His supply lines were lengthening. Winter was coming. And besides, he was anxious to get back to New York City where his American mistress awaited.

RAYS OF GLORY

There would be plenty of time to finish the rebels off in the spring. He halted his troops on the New Jersey side of the Delaware and ordered them into winter quarters, orders that included stationing fifteen hundred Hessians at New Jersey's future capital, at the time just a small river town—Trenton.

Three weeks later, early in the evening on Christmas Day, Hessian Colonel Johann Gottlieb Rall was enjoying a game of checkers when he heard the faint crack of musket fire.

Rumors were rife that George Washington was planning to attack Trenton, and Colonel Rall, the gruff old warhorse who commanded the garrison, wasn't about to be caught off guard. He had ordered his men to sleep in their uniforms and had been sending out patrols around the clock.

If an attack did come, Rall was certain of victory. To him, the rebels were nothing but "country clowns," so inept that he scoffed when one of his officers suggested fortifying the town. "Sheizer bey scheisz" (Shit on shit), he snapped. "Let them come. We want no trenches. We will come at them with the bayonet."

General Howe was just as confident. He knew from his loyalist spies what dire shape the rebels were in. Though Trenton was just across the Delaware from the rebel army, Howe had stationed thousands of troops just a day's march away. If the rebels were foolish enough to attack, they'd be walking into a hornet's nest.

Now, after Rall's well-drilled Hessians donned their brass miters and formed up, the colonel double-timed them in the direction of the musket fire, to a barrel maker's shop a mile from town where he'd posted some troops. A band of rebels had fired on the shop and then scampered off into the woods.

GLORY, GRIT AND GREATNESS

Above left: The British commander, General William Howe. *Above right*: A Hessian foot soldier, circa 1776.
(COURTESY OF THE NEW YORK HISTORICAL SOCIETY MUSEUM & LIBRARY.)

Rall balked at a suggestion that he reconnoiter the ferry landings upriver. The rumored attack, apparently nothing more than a raid on the barrel maker's shop, had seemingly come and gone, and a snowstorm, a real rager, was blowing in. After weeks of being on edge, for once he and his men could rest easy.

Rall's soldiers returned to their billets, where they spent the night drinking and caroling. Rall attended a dinner party at the home of a local merchant, where the colonel drank and played cards into the wee hours of the morning.

Late that night, a Pennsylvania farmer came pounding on the door with an urgent message. Denied entry by a servant, the farmer scribbled a hasty note and rode off. Rall pocketed the note without reading it.

The note was a warning that Washington's army had crossed the Delaware and was marching on Trenton.

Earlier that day on a chilly Christmas morning, at his headquarters in a farmhouse near the river, George Washington had been brooding over his attack plan.

He desperately needed a victory, but his crippled army—what was left of it—seemed incapable of winning one. He knew from *his* spies how vulnerable the Trenton garrison was, but if his plan failed, he and his men might be pinned against the river and cut to pieces.

A visitor saw the general scribbling something: "I observed him to play with his pen and ink upon several small pieces of paper. One of them by accident fell upon the floor near my feet. I was struck with the inscription upon it." It was the password Washington had chosen for the march on Trenton:

Victory or Death.

He planned to cross the Delaware on Christmas night and attack the town just before dawn on the twenty-sixth. By early Christmas evening—as Colonel Rall was hunched over his checkerboard—the plan was already unraveling. Ice floes on the river would keep the southern wing of Washington's force from joining the attack. That left only the northern wing, just twenty-four hundred men.

Unshod soldiers left bloody footprints in the snow en route to the crossing point, McConkey's Ferry, now part of Washington Crossing Historic Park. Getting the troops and horses and artillery over the ice-clogged river in the dark in a blizzard—"a perfect hurricane," said one soldier—became a ten-hour nightmare. Visibility was near zero. Gale-driven sleet clawed at hands and faces.

Above the howling storm, little could be heard but the booming voice of Colonel Henry Knox, the 280-pound behemoth in charge of the crossing. The former bookseller and future secretary of war was everywhere barking orders. One witness believed the effort would have failed but for Knox's "stentorian lungs."

Crowding into cargo boats resembling outsize canoes, soldiers stood in ankle-deep ice water, while rebel fishermen and sailors poled the boats across. Hurtling cakes of river ice rammed the gunnels. Men thrown into the water had to be fished out before they froze to death. Gunners slipped on icy mud as they manhandled their cannon onto ferry boats.

While waiting for their comrades, the first men across burned fence rails to keep from succumbing to the cold. "The wind and fire would cut them in two in a moment," sixteen-year-old fifer John Greenwood from Boston wrote in his journal. "When I turned my face toward the fire my back would be freezing.... By turning round and round I kept myself from perishing."

The crossing completed at last, Private Greenwood felt exhilarated: "The noise of the soldiers coming over and clearing away the ice, the rattling of cannon wheels on the frozen ground, and the cheerfulness of my fellow comrades encouraged me beyond expression, and big [a] coward as I acknowledge myself to be I felt great pleasure."

By then it was 4:00 a.m. and Trenton was still a ten-mile march away. Wet, freezing men, bleary-eyed from lack of sleep, trudged through the darkness by the light of torches flickering in the swirling snow. Two soldiers collapsed and died by the road. Numb with cold, Greenwood paused to sit on a stump and would have drifted off to sleep and died too if his sergeant hadn't roused him.

Horse-drawn cannon had to be unhitched and hauled across deep ravines. The soldiers straining at the ropes no doubt cursed the unwieldy guns, but they would come in handy later. Nineteen-year-old captain of artillery and future treasury secretary Alexander Hamilton walked beside one of his pieces as it rolled along, patting the barrel as though the gun were his child.

Top: Henry Knox by Gilbert Stuart. (COURTESY BOSTON MUSEUM OF FINE ART.)
Bottom: *Crossing the Delaware* by Emanuel Leutze. (METROPOLITAN MUSEUM OF ART.)

Rousted out of bed by his barking dogs, Dr. John Riker, an avid patriot, dressed quickly and joined the march. Before the day ended, he would save the life of a young soldier who would go on to become president of the United States.

Still five miles from Trenton as dawn was breaking, Washington held a hasty counsel of war on horseback. He would be attacking with less than half his troops and without the cover of darkness he'd counted on. It wasn't too late to call off the attack.

His orders were succinct: *Advance and charge.*

Later that morning, a squad of Hessians left the barrel maker's shop for a routine patrol. None too keen about being out in the storm, they stayed close to the shop, poked around a bit, and reported all quiet. At 8:00 a.m. the officer in charge, Lieutenant Andreas von Wiederholdt, decided to step outside. The stench of men who hadn't bathed was permeating the shop and he needed some fresh air. Through the blizzard he saw shadowy figures moving among the trees. Then he saw muzzle flashes. The figures were shooting at him. He rushed inside and shouted "Der feind!" (The enemy!)

Against all odds, despite attacking in broad daylight, despite all the Hessian patrols and loyalist spies and the warning note to Colonel Rall, Washington had achieved complete surprise. But his troops were still a mile from town. If Rall could assemble his men and mount a counterattack, he might yet win the day.

Two companies of Hessians from Rall's forward posts, including Wiederholdt and his men, tried to buy some time as they fell back, turning to fire from behind trees, split-rail fences, and houses, while the rebels came on with the fury of hounds chasing a fox.

As kettledrums sounded the alarm, Hessian soldiers came spilling out of their quarters into downtown Trenton. Dressed in his nightshirt, Colonel Rall poked his head out of a second-story window. "Don't you hear the firing?" his adjutant shouted. In a few minutes the old warhorse had donned his uniform and was galloping off to meet the attack.

Top: *Artillery Piece With Side Boxes,* dated 1776, by Charles Wilson Peale. (COURTESY OF THE AMERICAN PHILOSOPHICAL SOCIETY.) *Bottom*: Alexander Hamilton by John Trumbull. (LIBRARY OF CONGRESS.)

Whether because of the storm, or because Rall's troops were sluggish from their celebrating, or because the rebels, running on adrenaline and smelling blood, were moving so fast, the Americans were on the Hessians before they were ready, while they were still forming up.

Rather than fall back and summon help, Rall chose to slug it out in the center of town. It was a fatal mistake. The ragtag force of farmers and shopkeepers he'd called country clowns was suddenly performing like an elite corps. Rebel infantrymen broke away and cut Rall's escape routes, while others slipped into houses and shops and fired from windows. Rebel gunners rushed their cannon to the heights at the edge of town, catching the enemy in a withering crossfire.

Out in the open, their powder wet, their officers gunned down by sharpshooters, the Hessians were sitting ducks. Their fallen soon littered the streets, the wounded among them writhing in the snow in their blood-soaked tunics.

Colonel Knox detailed a squad to capture two guns that were shelling the heights. "Run for your lives!" yelled Sergeant Joseph White, a printer from Massachusetts, as the squad dashed toward the guns. The squad's leader, Captain William Washington, the general's cousin, was wounded in both hands, and the second-in-command, Lieutenant James Monroe, was hit by a bullet that passed through his shoulder and chest. The future U.S. president made it to an aid station where quick work by Dr. Riker kept him from bleeding to death. White raised his sword over a cowering Hessian gunner and shouted, "Run, you dog!" The two gun crews fled and White and his men wheeled one of the guns around and starting blasting away at the enemy.

Shot twice as he tried to rally his troops, Colonel Rall slid from his saddle, mortally wounded. Shown the warning note from his pocket, the dying colonel muttered, "If I had read this at Mr. Hunt's I wouldn't be here."

Meanwhile, his panicked men were fleeing down alleyways and into cellars. The remains of two regiments became trapped in an apple orchard east of town, where German-speaking rebels from the

Pennsylvania Dutch country called out for their surrender. Surrounded and beaten, the Hessians lowered their colors. After a futile attempt to ford a creek, a third regiment surrendered as well.

It was 9:45 a.m. The battle to save America had lasted less than two hours. Told of the surrender, George Washington broke into a rare grin.

He had reason to smile. At a cost of only four casualties, his men had killed or captured most of the garrison and taken a huge haul of equipment and supplies. Most prized at the moment were forty hogsheads of one-hundred-proof West Indies rum, the Hessian liquor supply. Soon, drunk rebels were stumbling around in brass caps snatched from the heads of the enemy. Washington ordered the rum poured out but didn't punish the offenders. If ever soldiers were entitled to a bender, his certainly were.

The Capture of the Hessians at Trenton, by John Trumbull, depicts a dying (and hatless) Colonel Rall standing beside Washington's horse. In fact, the two commanders never met up after the battle. Lieutenant James Monroe, the future president, lies wounded behind Rall's right hand. (YALE UNIVERSITY ART GALLERY.)

The rebel army's march back to Pennsylvania was as grueling as the one on Christmas night. Men perished along the way from exposure and exhaustion. Upon reaching camp, Sergeant White collapsed. "I being weary laid down upon the snow," he wrote. "The heat of my body melted the snow and I sunk down to the ground."

White and his comrades had been marching and fighting almost nonstop for more than two days. They had performed superbly, won a great victory, and earned a long rest.

They wouldn't get it. On December 30, Washington led his men back across the Delaware River and dug in behind Assunpink Creek just south of Trenton, a risky move given that most of his army's enlistments were due to expire on New Year's Day. With the enemy preparing to march on the rebel camp, Washington went among his men, imploring them to re-enlist.

In a Connecticut regiment, down to two hundred ragged, shivering souls, no one stepped forward to volunteer. Washington wheeled his horse around and rode along the ranks to make a last-ditch plea:

> My brave fellows, you have done all I asked you to do and more, but your country is at stake, your wives, your houses, and all that you hold dear. You have worn yourselves out with fatigues and hardships but we know not how to spare you. . . . [This] is emphatically the crisis which is to decide our destiny.

After an awkward silence, a few men edged forward. Then a few more, followed by nearly all two hundred. None of them had any illusions about what lay ahead. Half the men in the regiment would never see their homes again.

In similar scenes throughout the day, most of Washington's troops agreed to stay. For the time being at least, his force would remain intact.

Unlike his superior, General Howe, British General Charles Cornwallis was no procrastinator. By daybreak on January 2nd, at the head of eight thousand men, he was en route to attack the rebel army. The Hessians in his force had been ordered, under threat of the lash, to kill any rebel who surrendered.

A downpour the night before had turned the road to Trenton into mush, slowing Cornwallis's twelve-mile march from Princeton. His column was barely underway when gunfire rang out from a patch of woods up ahead. The shooters were among the world's deadliest marksmen, moccasin-wearing woodsmen from the Pennsylvania backcountry armed with long rifles, part of a delaying force sent by Washington, led by Irish-born physician Colonel Edward Hand.

All day long, Hand's men peppered the slogging column with rifle fire from beyond the range of the enemy's smooth-bore muskets, then melted into the trees, scattering British and Hessian dead all along the route and costing Cornwallis precious time. As the enemy entered Trenton, Hand's men fired from the same houses and shops the Americans had used for cover a week earlier. His delaying mission done, he and his men slipped across the Assunpink Creek bridge and rejoined the rebel army.

By now it was dusk. Sitting atop his horse over a lengthening shadow, Cornwallis scanned the rebel works across the creek with his telescope and mulled over what to do next. Thanks to the New Jersey mud and Colonel Hand's sharpshooters, there wasn't enough daylight left to ford the creek upstream and roll up Washington's flank. As aggressive as General Howe was cautious, Cornwallis ordered an immediate frontal assault. Surely the rebels would buckle, as they had so often before, after which he would crush them against the Delaware River.

GLORY, GRIT AND GREATNESS

Above left: Colonel Edward Hand. (NEW YORK PUBLIC LIBRARY.)
Above right: General Charles Cornwallis by Thomas Gainsborough
(NATIONAL PORTRAIT GALLERY, LONDON.)

All eyes turned to a stone bridge—not much wider than a horse-drawn carriage—in front of the rebel trenches. Grenadiers massed shoulder-to-shoulder came charging across. The rebels held their fire until the last second, then opened up with everything they had. Amid an ear-splitting blast, a thousand small arms, along with cannon loaded with canister, swept the bridge. Four times the grenadiers charged, and were stopped each time literally dead in their tracks, as cheer after cheer erupted along the American line.

The cheering stopped when the smoke cleared and the rebels saw the horror they had wrought. British corpses in red uniforms packed the bridge so tightly, said Sergeant Joseph White, that it seemed awash in blood. On the ground beyond, another rebel said, the enemy dead were as plentiful as sheaves on a wheat field.

Although the Americans had won another victory, Cornwallis had Washington right where he wanted him. "We've got the old fox safe now," the British general assured his staff that night. "We'll go over and bag him in the morning."

The fox had other ideas. The next morning Cornwallis awoke to find Washington and his army had vanished. As Cornwallis pondered this startling news, cannon rumbled in the distance. Leaving their campfires lit, muffling their cannon wheels with rags, the Americans had slipped away in the dead of night and were attacking Cornwallis's rear guard in Princeton. On Christmas night, a blizzard had masked Washington's attack on Trenton; a week later rain had turned Cornwallis's march into a muddy slog, and now a sudden overnight freeze had hardened the roads again, allowing Washington's army to escape. For the third time in a week, the weather gods had smiled on the rebels.

Cornwallis rushed his troops back to Princeton, but by the time he got there it was too late. The Americans had routed the garrison and captured another large haul of supplies. Before the British could catch them, Washington and his men slipped away again, to Morristown, New Jersey, safely behind the Watchung Mountains.

News of the rebel victories sent shock waves all the way to London, where even phony casualty lists and censored reports couldn't hide the truth. The British had suffered a major disaster. Public support for the war began to wane. General Howe ordered a humiliating retreat, abandoning New Jersey and retracing his steps all the way back to New York.

The British would return, of course, with an army as mighty as before. In the five years of fighting that remained, they would win many more battles and would even take Philadelphia, but they would never again come as close to victory as they had in December 1776. The American Revolution would never again teeter on the brink of collapse. "All our hopes were blasted," colonial secretary Lord George Germain lamented to Parliament, "by that unhappy affair at Trenton."

In the decades to come, the Americans who fought there would be lionized by a grateful country, but for now they were just glad to have the fighting over for a while. They wrapped themselves in captured blankets, settled into the log huts they'd built at their winter camp in Morristown, and drifted off to sleep.

CHAPTER 4

THE PERFECT STICK

Calvin Coolidge

Calvin Coolidge, someone once said, looked like he'd been weaned on a sour pickle.

America's thirtieth president was known for his tight-lipped stare and his even tighter-lipped personality. A hostess once bet him she could get more than two words out of him. He replied, "You lose." Even his wife Grace, the sparkling, vivacious woman who loved him, called him "a perfect stick."

Yet, the gruff, funny-looking little man, who, when he did talk, sounded like a quacking duck, became one of America's most popular presidents.

Crushed by a personal tragedy, "Silent Cal" walked away from the spotlight and into obscurity, where he remains today, the greatest president no one remembers, the socially awkward misfit who led America through the Roaring Twenties—who, at heart, would always be a shy little boy from the Vermont hills.

GLORY, GRIT AND GREATNESS

At dinner parties, Silent Cal usually sat staring at his plate, answering in monosyllables when someone tried to strike up a conversation. But let the conversation turn to Vermont, and to everyone's surprise he would perk up and become practically gabby.

Stunning fall foliage aside, Vermont's rugged Green Mountains, where he was born and raised, can be a hard place to live. Many of the pioneers who migrated there moved west after a few years. The ones who stayed, scratching a living from the rocky soil for generations, were hard-headed and tight-fisted, traits for which the future president would become well known.

Descended from English Puritans who had lived in the mountains since the American Revolution, he was born John Calvin Coolidge on July 4, 1872—the only president born on Independence Day—in the backwoods hamlet of Plymouth Notch, a smattering of tiny hill farms arrayed around a church, a schoolhouse, and a general store in the middle of a maple forest, twelve miles from the nearest train station.

Calvin's dynamo of a father, John Coolidge, was a jack-of-all-trades, a man of boundless energy who did stints as the town's constable, justice of the peace, and state legislator; ran the general store, owned a small foundry, sold insurance, taught Sunday school, and manufactured cheese, besides being a skilled brick layer, carpenter, cobbler, and veterinarian.

Calvin idolized his father, and they maintained a close relationship, nurtured by frequent letters, until John's death in his eighties. "I cannot recall that I ever knew of him doing a wrong thing," Calvin wrote in his autobiography. "I was exceedingly anxious to grow up to be like him."

Calvin's mother, Victoria Moor Coolidge, provided warmth and affection, as well as the tender streak under her son's aloof exterior.

THE PERFECT STICK

"Whatever was grand and beautiful in form attracted her," he wrote. "It seemed as though the rich green tints of the foliage and the blossoms of the flowers came for her in springtime, and in autumn it was for her that the mountainsides were struck with crimson and gold."

An invalid, Victoria died when Calvin was twelve, which devastated him. As much as he adored his wife, he would never be as close to anyone as he'd been to his mother. Decades later, he still reminisced about her delicate beauty and love of flowers. "He seemed to remember every day he spent with her," a friend recalled. The first thing he did when he moved into the White House was put her picture on his desk. When she died, he withdrew into a shell from which he never fully emerged.

John Coolidge's business ventures brought in good money, and by "Notch" standards, the family prospered. But no one in the tiny, isolated town led a life of luxury. There were no gas lamps, running water, or coal-burning furnaces. Calvin's first trousers were a pair that had belonged to his grandfather. The Coolidges spent the winter living in their kitchen to be near the wood-burning stove, and washed up from a basin of ice water. During heavy snowfalls, the town was cut off from the outside world. Survival depended on storing up enough food to last until spring.

Young Calvin, the only boy in a family of two children, pitched hay and dug potatoes on his family's two-acre plot, peddled apples and cordwood to the neighbors, and tapped maple trees for syrup. (His father once bragged that his son could get more sap out of a maple than any boy in town.) Workdays in Plymouth Notch lasted from dawn to dusk. Hard work and self-reliance were considered virtues, while sloth was considered immoral, beliefs that would become part of Calvin's politics.

His grammar school teacher remembered an unexceptional boy with little better than average grades. Not until he went away to high school did he show flashes of promise. At Vermont's Black River Academy, where he often rose at 3:00 a.m. to get in some extra studying, he mastered a daunting curriculum that included classic literature, Latin, Greek, and French.

At the school's social events, Calvin stood off by himself, too shy to dance with the girls. Yet he became an articulate public speaker and was chosen to give the traditional farewell address to the graduating class. As a senior, as part of his final exam, he spoke on a topic fit for a doctoral student—*The influence of oratory in the formation of public opinion and in the great movements of history.* A local newspaper called his speech "masterly."

To help pay his room and board, he worked part-time in a toy factory. "The man who builds a factory builds a temple," Coolidge the statesman would later say. "The man who works there worships there, and to each is due, not scorn and blame, but reverence and praise."

At Amherst College, a small, liberal arts school in Massachusetts, he became known as a "grind," a student who spent most of his time studying. "A drabber more colorless boy I never knew than Calvin Coolidge when he came to Amherst," said a classmate. To spice up his image, Calvin started wearing a derby and carrying a cane. It didn't work. No fraternity would have him. Despite his hard work, his freshman grades were subpar.

"I don't seem to get acquainted very fast," he lamented in a letter to his father. "Many men in my class have the strength, preparation, inclination and ability to do much more than myself."

Calvin seriously considered dropping out of school and returning to Plymouth Notch to become a storekeeper like his father. Instead, he persevered, upping his grades enough to graduate cum laude. An essay he wrote on the American Revolution won a national award. He won a place in the "Hyde Fifteen," the top fifteen orators

Clockwise from upper left: The future president age 7. (COURTESY OF THE FORBES LIBRARY, NORTHAMPTON, MASSACHUSETTS.) At Amherst College. (COURTESY OF THE VERMONT HISTORICAL SOCIETY.) Calvin's parents John Coolidge (LIBRARY OF CONGRESS) and Victoria Moor Coolidge. (CALVIN COOLIDGE PRESIDENTIAL FOUNDATION.)

of his class, and was at last accepted into a fraternity. Admired for his dry wit, he was chosen to give the commencement roast on graduation day. His speech, delivered in deadpan to howls of laughter, was a smash.

Among those howling the loudest was a local attorney, an Amherst grad named Henry Field, who would soon hire the young man and start him on his improbable rise from city councilman to president of the United States.

In his senior year, Coolidge wrote his father asking for permission to study law. The son said he wanted to be of use to the world more than he wanted to make money. After graduating, Coolidge became an apprentice at Hammond and Field, a law firm in Northampton, Massachusetts, a city of fifteen-thousand across the Connecticut river from Amherst. Henry Field remembered Coolidge as an "inscrutable little devil," smart and diligent, yet also prone to sit in his chair and stare out the window deep in thought.

Coolidge caught the political bug from Field and his partner, John Hammond, who were active in local politics. Like nearly everyone from Vermont in those days, Coolidge was a Republican. In 1898, a year after entering the bar, the twenty-six-year-old tossed his hat in the ring, declaring his candidacy for the Northampton City Council.

Coolidge's personality seemed better suited to running a general store than running for office. "The last place . . . I would have expected him to succeed was politics," recalled a college classmate. "He lacked small talk, and he was never known . . . to slap a man on the back. He rarely laughed. He was anything but a mixer."

Nor would Coolidge's looks win him any votes. Five feet nine inches tall with narrow shoulders and spindly legs, he looked rather

THE PERFECT STICK

frail. His facial features—a pointy chin, squinty eyes, a beak for a nose, and an outsize forehead that loomed beneath a receding hairline—fell somewhere between homely and comical.

Yet, over the next twenty years, he ran for office sixteen times in Massachusetts and lost only once, an early race for school board when a fellow Republican split the vote. As Coolidge climbed the political ladder—from city councilman, to assemblyman, to mayor, to state senator, to lieutenant governor, and finally to governor—he faced ever more formidable opponents and bested them all.

In the early years, he spent countless hours walking precincts and knocking on doors, a major ordeal for someone so shy. Thankfully, he didn't have to chat a lot. A handshake, a few words about the campaign, a quick plea for support ("I want your vote. I need it. I shall appreciate it."), and he'd be on his way to the next house.

What people saw in the awkward candidate who "quack[ed] through his nose" in squeaky Vermontese, as one wag put it, is anyone's guess. Something in his quiet manner must have impressed them. Whatever it was, Coolidge became, said the *Boston Evening Transcript*, "one of the biggest vote getters" in Massachusetts history, including votes from many "Coolidge Democrats," the Irish immigrants who were turning the state into a Democratic bastion. In his state senate race in 1913, a year that saw the state elect its first Irish Catholic governor, and Democrats sweep to victory statewide, Coolidge was elected handily.

Hopping on the bandwagon, Northampton's *Hampshire Gazette* chirped:

Now for Coolidge we will sing,
Honest Cal must have his fling
Loud for him our praises ring
Cal was born to be a king.

Presiding over the state senate, Coolidge was known for the gruff way he responded to bad ideas—"No."—a forthright style that caught the eye of fifty-eight-year-old Boston department store tycoon Frank Waterman Stearns.

Soon, the avuncular Stearns was telling anyone who would listen that the obscure state senator from Northampton should be president of the United States. Putting Coolidge in the White House became Stearns's passion. Asking nothing in return, driven by a devotion to his protégé no one else fully understood, Stearns managed Coolidge's campaigns, raised money for him, and donated a large chunk of his own fortune to the cause.

When Coolidge ran for lieutenant governor in 1915, he hitched up a buggy and drove around the state, touring with GOP gubernatorial candidate Samuel McCall and giving as many as fifteen speeches a day. "McCall could fill any hall in Massachusetts," someone quipped, "and Coolidge could empty it." He made up for his stiff manner on the stump, or tried to, with lofty prose:

> Let the laws of Massachusetts proclaim to her humblest citizen, performing the most menial task, the recognition of his manhood, the recognition that all men are peers, the humblest with the most exalted, the recognition that all work is glorified. Such is the path to equality before the law.

The voters, the only critics who mattered, liked what they heard. "In gestures and flights of oratory," reported *The Boston Herald*, Coolidge "nowhere near measured up to [his opponent] the picturesque Mr. Barry, but he won the confidence of the people." On election day, Coolidge outshone both the flamboyant McCall and the even more flamboyant Barry. McCall won only narrowly, while Coolidge beat Barry in a landslide.

THE PERFECT STICK

Coolidge's string of easy victories ended when he barely won the 1918 governor's race. His opponent, businessman Richard Long, ran an aggressive, well-funded campaign that was, according to one Coolidge backer, one of the dirtiest in memory.

Coolidge could have hit back—there were allegations that Long had overcharged the military for some supplies—but chose not to. Throughout his career, Coolidge refused to personally attack his opponents no matter what they said about him. He said he would rather lose an election than engage in mudslinging.

He would run for office three more times, winning every race, the last stages of what one historian has called "the steadiest political rise in American history." An old college friend once asked Coolidge the secret of his success. "I guess fortune did it," he shrugged. Pressed for a better answer, he deadpanned a reply in his thickest Green Mountain drawl, "Wal mebbe I did nudge fortune some."

Governor Coolidge. (LIBRARY OF CONGRESS.)

GLORY, GRIT AND GREATNESS

Focused on politics, Coolidge put aside any thoughts of marriage. He might have been the only U.S. president besides James Buchanan who never married, if his curtains had been drawn the day Grace Goodhue happened by.

One morning, she saw Coolidge shaving in the window of his second-story Northampton apartment, standing there in his long johns with a derby perched on his head, his chin swathed in lather. She glanced up and burst out laughing; he looked down and was smitten. He arranged for a meeting and before long, the couple was keeping company.

The daughter of a steamboat inspector, Grace had a degree from Vermont University and would eventually become the first first lady to have graduated college. When she met Calvin, she was working as a teacher of the deaf, a career she abandoned after they married. Grace's mother, who disapproved of him, and Grace's friends, who gossiped about him, thought she was too good for him.

They seemed to have a point. As Grace gracefully put it, she and Calvin had "vastly different temperaments and tastes." She was a vivacious twenty-six-year-old, a willowy brunette with a beaming smile and an eye for fashion. He was a thirty-two-year-old stick-in-the-mud.

At parties, while she was off mingling, he would plant himself somewhere away from the crowd. A guest whispered, "That man standing in the corner, is that one of Grace's [deaf] pupils?" When the couple visited one of Grace's college friends, Calvin sat there with a blank stare without saying a word. Afterward, Grace chided him for being a perfect stick. Eventually, she learned to tolerate such behavior. Once, while she was first lady, when someone remarked how extroverted she was, she laughed and said, "Well, I have to talk for two."

THE PERFECT STICK

His pet name for her was Momma. She called him Papa. She loved him in spite of himself and he adored her. Someone who knew them well described Grace as the "sunshine and joy" of Calvin's life, "his rest when tired, his solace in times of trouble." At her approach, "light broke across his grave face."

He wrote her lovestruck letters ("you are like the morning in my own Green Hills") and delighted in surprising her with small gifts, often frilly Victorian hats and dresses that had been out of fashion for years and that sat, unworn, in her closet. "We thought we were made for each other," he would write. "For almost a quarter century she has borne with my infirmities and I have rejoiced in her graces."

Whatever Grace saw in Calvin, she didn't marry him for his money. At the time of their wedding, a year before his first race for the state legislature, he was still a struggling lawyer. With the birth

Grace Coolidge, née Goodhue.
(COURTESY OF THE VERMONT HISTORICAL SOCIETY.)

of the couple's first son, John, money became tight, and even tighter when another son, Calvin Jr., arrived. The Coolidges lived in a Northampton duplex they rented for twenty-seven dollars a month. Grace economized by buying second-hand bedding and silverware from a hotel.

She could have earned extra money teaching, but, like most husbands in that era, Coolidge expected his wife to be a full-time homemaker, a role that Grace, a product of those same times, willingly accepted. After the wedding, he'd handed her several dozen pairs of socks to darn. She asked him if he'd married her just to have them mended. "No," he replied. "But I find it mighty handy." Rather than tell him what he could do with his socks, she dutifully saw to the chore.

Even when Coolidge was making $10,000 a year as governor, a tidy sum in those days, he kept a tight rein on the family purse. In the days when governors had to provide their own housing, the Coolidges lived in a cramped two-dollar-a-day apartment. On trips around Boston, Governor Coolidge rode the trolley.

Once, without telling him, Grace spent eight dollars on a copy of *Our Family Physician*, a book of home remedies. She set the book on the coffee table and later found an inscription on the flyleaf: "Don't see any recipe here for curing suckers! Calvin Coolidge."

The event that turned Coolidge from the little-known governor of a small state into a national figure was the 1919 Boston Police strike.

Spurred by soaring postwar inflation and stagnant wages, four million American workers—steel workers, transit workers, coal miners, actors, house painters, shipbuilders, bricklayers, and countless others—walked off the job that year in thirty-six hundred strikes.

THE PERFECT STICK

Sixty-five thousand workers struck in Seattle, virtually shutting down the city and bringing it to the brink of martial law. Socialists and police clashed during a workers' march in Cleveland, leaving two dead and dozens injured. Anarchists sent bombs to businessmen and politicians, leading to the jailing and deportation of radicals in the infamous Red Scare. Agitators stirred up workers with anticapitalist handbills and speeches, giving rise to fears of a Soviet-style revolution.

Labor had become America's hot-button issue, as explosive as the Vietnam War would be fifty years later.

Amid the turmoil, the Boston Police unionized. They had legitimate grievances—meager pay, long hours, rat-infested station houses—but there was also a long-standing department rule against police unions, a rule every officer had agreed to.

The city's hard-line police commissioner, Republican Edwin Curtis, threatened to suspend the union leaders. The police responded by threatening to strike. Boston's Central Labor Union, representing eighty thousand local workers, pledged to support the police. The city seemed headed for a disaster.

For Governor Coolidge, in the midst of a tough re-election campaign in a state with a strong pro-labor vote, the safe course was to fire Curtis and clear the way for a compromise. Instead, Coolidge issued a statement giving Curtis his support. Friends warned Coolidge that refusing to seek a compromise would cost him the election. He said it wasn't necessary for him to be re-elected, then turned and stared out the window until his visitors got up and left. The stubborn Vermonter had made up his mind.

Curtis went ahead with the suspensions, and the next day eleven hundred officers, nearly three-quarters of the police force, walked off the job to the cheers of a pro-union crowd.

Two nights of rioting followed. Roving mobs, some incited by striking policemen, smashed windows and looted shops. A streetcar was waylaid and its passengers pelted with rocks. The conductor was shot in the leg. A group of volunteer cops being hit with rocks and bottles fired on a mob, killing three people.

By the third night, the state militia having arrived on Coolidge's orders, the streets had quieted down. The violence might have been avoided if Coolidge hadn't delayed calling out the militia, a delay he later admitted was a mistake.

The strike, meanwhile, was losing steam. The local press, which had been calling for a compromise, was now condemning the police. President Woodrow Wilson chimed in, calling the strike "a crime against civilization" that had left Boston "at the mercy of an army of thugs." Even the pro-labor *New Republic* accused the strikers of shirking their duty. The Central Labor Union backed down, and in the wake of the violence, public opinion turned against the officers.

Realizing they were beaten, the police abandoned the strike and asked to return to work, a request Commissioner Curtis denied with Governor Coolidge's backing. Ironically, some of the ex-cops would find new jobs with Coolidge's help.

The strike, which was front-page news throughout the country, had thrust Coolidge into the national spotlight. The stage was now set to make him a political hero. When Samuel Gompers, the head of the American Federation of Labor, sent him a telegram denouncing Curtis, Coolidge sent a sharply worded reply, including a line that made headlines: "There is no right to strike against the public safety by anybody, anywhere, any time."

"BAY STATE GOVERNOR FIRM," trumpeted the *New York Times* the next morning. Editorials extolled Coolidge's toughness. Overnight he became the champion of those who saw the Boston Police as no better than deserters. Letters of support poured into his office. In November he won reelection by a landslide. He was even being mentioned as a possible nominee for president.

The shy boy from the backwoods of Vermont was now a major player in national politics.

Considering they were about to nominate a sure winner, Republicans had every right to feel jubilant as they gathered at the Chicago Coliseum for their 1920 national convention. But it was hard to work up any enthusiasm in the godawful heat.

Though summer was still a week away, sunshine streaming through the skylights in the roof, past rows of sunlit American flags hanging from the rafters, had turned the coliseum into a giant hot house, driving the temperature up to nearly a hundred degrees. As the week wore on, delegates shed their coats, ties, even their shoes, in a vain attempt to cool off.

As if the heat weren't bad enough, the place also stank. Journalist H. L. Mencken, covering the convention for the *Baltimore Sun*, described the coliseum as "a foul pen" used for "prize fights, dog shows and a third-rate circus" that smelled of "pugs, kennels and elephants."

After eight years of a Democrat in the White House, and with the economy struggling, the Republicans were heavy favorites to regain the presidency in November. The real battle for the White House would be fought at the GOP convention. Coolidge was a dark horse in a wide-open race for the nomination. Bookies had set the odds against him at eight to one.

Four sweaty days, seven mind-numbing hours of nominating speeches, and six ballots later, the delegates were deadlocked, their votes spread among fifteen candidates, opening the way to a brokered convention. That night a handful of senators, the GOP's kingmakers, gathered in a smoke-filled Chicago hotel suite to pick the next president.

Coolidge, meanwhile, was literally playing it cool, sitting out the convention in Boston—a balmy seventy degrees—and tending to his duties as governor, content to let his friend and patron, Frank Stearns, do his campaigning for him.

Stearns made a serious run at getting his man nominated, sending each delegate a pocket-size booklet of quotes from Coolidge's speeches with the delegate's name printed in gold on the cover ("as neat and effective a piece of political publicity as I've ever seen," said one politico), and choosing Shakespearean actress Alexandra Carlisle Pfeiffer to second Coolidge's nomination. Ms. Pfeiffer's performance was a corker. Her speech drew thunderous applause. Coolidge the dark horse had become the delegates' sentimental favorite.

He might have walked away with the nomination were it not for the bosses in the hotel suite. In the wee hours of the morning the bleary-eyed group finally settled on a fellow senator, Warren Harding of Ohio, "the best of second-raters," as one boss called him, to be the nominee.

The delegates weren't too pleased with having Harding foisted on them. They nominated him the next day, but only after ten more ballots. On getting the news of Harding's victory by phone, Coolidge went for a stroll around Boston Common. On returning, he lit a cigar and was settling in for a quiet evening with Grace when the phone rang again. The caller informed him he'd just been nominated for vice president.

It had been a wild finish. Choosing Harding's running mate had been the convention's last order of business. Everyone was anxious to wrap things up and go home. Many delegates had already left. Some of the bosses got together under the speakers' platform and picked one of their friends, Wisconsin Senator Irvine Lenroot, to be the nominee.

Out on the floor, Oregon delegate Wallace McCamant, a rebel with a bullhorn voice, was waving his arms, demanding to be heard. Thinking McCamant was going to second Lenroot's nomination, the chairman recognized him. McCamant stood on a chair and in his booming voice nominated "a man who is sterling in his Americanism and stands for all that the Republican Party holds dear . . . *Governor Calvin Coolidge of Massachusetts!*"

The floor erupted. The delegates unleashed all their pent-up anger—from the heat and stench, from all the tedious speeches, and from having first Harding and now Lenroot rammed down their throats—in a burst of acclaim for Coolidge. Delegates were shouting over each other to second his nomination. Soon the hall was ringing with *WE WANT COOLIDGE!* Illinois Senator Medill McCormick, a GOP bigwig, rushed to the podium to halt the stampede, but it was too late. Coolidge was nominated overwhelmingly, and would soon be a heartbeat away from the presidency.

As the convention wound down, the coliseum was abuzz with talk of the "Coolidge luck." H. L. Mencken overheard a fellow reporter offering to bet all comers that Harding would die in office and Coolidge would succeed him. "I am simply telling you what I know," the reporter said. "I know Cal Coolidge inside and out. He is the luckiest son of a bitch in the whole world."

By the 1920 presidential race, which Harding won in a landslide, Coolidge was forty-eight years old and fed up with campaigning. A reporter who asked him if he was enjoying the campaign got barked at: "I don't like it. I don't like to speak. It's all nonsense. I'd much rather be at home doing my work."

GLORY, GRIT AND GREATNESS

Top: 1920 Republican National Convention. *Bottom*: Frank Waterman Stearns. (BOTH PHOTOS: LIBRARY OF CONGRESS.)

As vice president, Coolidge was best known for giving the shortest inauguration speech ever—five hundred words—and for turning down a chance to occupy a Washington, D.C. mansion owned by a rich Republican donor. Instead, Coolidge and his family moved into a small apartment in a D.C. hotel, where they lived in semi-obscurity. After a fire alarm, as the Coolidges headed back to their apartment, the fire marshal stopped them and asked Coolidge who he was.

THE PERFECT STICK

"Vice president," he replied. The marshal started to let them pass then stopped them again with another question: "Vice president of what?"

Two and a half years into Harding's presidency, amid talk that Coolidge would be dropped from the ticket in the next election—payback for beating out Lenroot—President Harding, mired in the Teapot Dome scandal, dropped dead, probably of a heart attack.

At the time, Calvin and Grace were spending a few days in Plymouth Notch and couldn't be reached by phone. Word of Harding's death finally came via a messenger in the wee hours of the morning. The only notary public at hand was Calvin's father. At 2:47 a.m. on August 3, 1923, by the light of a kerosene lamp in the parlor of Calvin's boyhood home, John Coolidge administered the oath of office to his son.

Years later he revealed what flashed through his mind when he learned he was headed for the White House—*I thought I could swing it.*

A GOP senator wasn't so sure. "Oh my God!" he gasped when he heard the news. "Coolidge is president."

Taxes slashed, budgets balanced, rock-bottom unemployment, a booming economy; during Coolidge's presidency the twenties roared and the good times rolled.

The shy hillbilly from Vermont became, said the *New York Times*, "one of the shrewdest" presidents America ever had. Hardly the dimwitted tool of the rich portrayed in history books, he was smart and incorruptible, an avid reader of the classics, fluent in Latin, who translated works by Dante and Cicero, and told donors not to expect any favors.

One historian claimed Coolidge "slept away" his presidency. He did sleep a lot, eight hours a night plus a midday nap. "Nero fiddled,"

Mencken quipped, "but Coolidge only snored." Suffering from the heart disease that would prove fatal within a decade, he needed plenty of rest.

He got his work done by sticking to a strict routine: up at 6:30 a.m. followed by a walk around the White House grounds; a quick breakfast at 8:00—fruit, hot cereal, a roll, and half a cup of coffee, along with a strip of bacon that usually went to his pet collies, Prudence Prim and Rob Roy; meetings and speech writing until 12:30; shaking hands with the public until 1:00; then lunch, usually with White House guests; a nap of an hour or two; reading official documents in the late afternoon and early evening; another walk; dinner at 7:00; dictation to his stenographer from 7:45 to 10:00; then to bed.

Unlike Harding, who had let his work pile up, Coolidge finished his before bed and awoke to a clean desk.

He refused to install a telephone in his office, the better to avoid chitchat. Anyone who wanted his ear had to come to his office and face excruciating silence. In a story reported in the *New Yorker*, GOP activist Ruth Hannah McCormick sent a group of Polish Americans to see the president about getting a judgeship for a fellow Pole. During the meeting, Coolidge sat stone-faced in his chair, saying nothing, until eventually he and his bewildered visitors were all staring silently at the floor. "Mighty fine carpet," he finally said. "New one. Cost a lot of money. She wore the old one out trying to get you a judge." Realizing they were wasting their time, the group gave up and left.

Welcome visitors received the same treatment. Coolidge would invite Stearns to his office, then say nothing as the two men sat staring out the window. Once, when Stearns got up to leave after an hour, Coolidge said, "Stay a while longer."

In contrast, he talked freely to the nation through radio, the era's revolutionary new medium, opening up what journalist William Allen

THE PERFECT STICK

President Coolidge in the Oval Office. (LIBRARY OF CONGRESS.)

White called "a channel of communication between his soul and the soul of the American people." No spellbinder, Coolidge won people's trust with his nasally Vermont twang and plain-spoken style.

Back when voters weren't as easily offended as they are now, his bluntness was an asset. He said things that today would be political suicide, such as: "If any man is out of a job, it is his own fault," and "The normal must care for themselves." He was at his flintiest, however, when attacking excessive taxes, calling them "legalized larceny," the government that imposed them "an instrument of tyranny."

Later presidents would wage a war on poverty, a war on drugs, and a war on terror. Coolidge waged war on the federal income tax, pushing through the GOP-controlled Congress bills that slashed taxes for rich and poor alike.

In the wake of a postwar slump, surplus cash from the tax cuts ignited the "Coolidge prosperity," an economic boom of historic proportions. By 1929, the year he left office, America's factories were churning out forty percent of the goods produced by the entire world.

Wages rose, prices fell and unemployment plummeted. Sales of radios, refrigerators, washing machines, vacuum cleaners, and cars soared, as ordinary Americans bought luxuries previously reserved for the rich. Americans bought homes in record numbers. Meanwhile, the workday was shrinking from ten or twelve hours to eight.

With workers more content, the labor trouble of a few years earlier eased. The number of striking workers fell from four million in 1919 to less than four hundred thousand in 1927. "As long as men have enough money to buy a second-hand Ford and tires and gasoline," a labor leader lamented, "they'll be out on the road and paying no attention to union meetings."

One historian was hard-pressed to find "any aspect of [American] culture in which the 1920's did not mark spectacular advances." Education spending and literacy rose dramatically, as did patent applications and air travel. As profits increased and the rich got richer, philanthropic spending rose. Labor-saving devices and the shorter work week gave rise to more leisure time. Americans pocketed their hip flasks, cranked up their Tin Lizzies, motored headlong into the Jazz Age, and never looked back.

They flocked to movie theaters to hear Al Jolson crooning on screen in *The Jazz Singer*, the first talkie, and packed Yankee Stadium to watch Babe Ruth blast tape-measure homers. When Charles Lindbergh flew the Atlantic in June 1927, America went wild, celebrating the country's most triumphant moment between the world wars.

All this happened under Harding and Coolidge, mostly under Silent Cal. Not that there weren't problems. Hard times persisted for farmers, who'd been hit especially hard by the postwar slump, and too much of the country's spending binge was done on credit, which spelled trouble down the road. But overall, the Roaring Twenties were an extraordinary decade, a time of progress and prosperity, when the American dream came true for more Americans than ever. Coolidge's tax cuts provided a spark that helped make the boom times possible.

Despite cutting taxes, he balanced the federal budget each year he was in office, for five years running, a feat no other twentieth-century president would match. "I am for economy," the tight-fisted Vermonter declared. "After that I am for more economy." Belt-tightening in the bureaucracy became the order of the day. Federal workers were issued one pencil at a time, receiving another when they returned the stub.

Coolidge vetoed price supports for farmers, and veterans' bonuses, the latter of which Congress overrode, vetoes that, even back when voters were more fiscally conservative, took political courage. When the Mississippi River had one of its worst floods ever, causing mass homelessness, Coolidge cut by two-thirds the $1.4 billion Congress wanted to spend for flood relief, an amount that would have wiped out the federal budget surplus that year. Although he sympathized with the victims, he believed private charities and the states hit by the flood should pay the lion's share of the aid.

He also correctly predicted that, by stimulating the economy, cutting taxes would actually *increase* federal revenues, a theory that would later gain fame as supply-side economics. Due to the spending restraints and extra revenue, the national debt, which had exploded in the wake of America's involvement in World War I, fell sharply under Coolidge.

Behind the scenes, the high-minded statesman became a starry-eyed kid. Coolidge would "almost tiptoe" around the White House, "touching things and half smiling to himself," his bodyguard and fishing buddy, Secret Service Agent Edmund Starling, wrote in his memoirs. "It was as if he were a small boy whose daydream of becoming king had suddenly been made real by the stroke of a magic wand."

The president pouted when Starling caught more fish than he did, and he also played silly pranks on the White House staff. His favorite was to press the buzzer that signaled he was returning to the White House from his office, then slip out a side door for a stroll, while servants and secret servicemen scurried around preparing for his arrival.

A junk-food junkie, Coolidge binged on sausage, cheese, cake, crackers, and candy, a habit that no doubt worsened his chronically sour stomach and heart trouble. Starling described the president sneaking into the White House banquet hall before a reception "to steal and eat the choice cakes."

In his lighter moods, he evinced a bone-dry wit. Scoring with a zinger, he would never crack a smile, the only hint of mirth being a twinkle in his blue eyes. Like the time he and Grace were getting separate tours of a chicken farm. Grace asked her guide how often a rooster copulates. Dozens of times a day was the answer. "Tell that to the president," Grace chortled. When told, Calvin asked, "Same hen every time?"

Even Coolidge's friends had to admit he was an odd duck. Grace, on the other hand, was unreservedly beloved. She "was never beautiful; she wasn't even what you'd call pretty," said an admirer. "She just had a pleasant average face and a slender average figure, but her genuine interest in you, which shone through her warm dark eyes, and her kindness, seemed to cast a sort of glow wherever she went."

Motivated by sexism, or maybe jealousy, Calvin forbade Grace from granting interviews for personal news stories. At his insistence, she confined herself to charity work and acting as hostess during White House social events. In public she walked behind the president, staying squarely in his shadow.

Yet she still made her mark. Just as Jackie Kennedy would forty years later, Grace influenced women's fashions (her pleated skirts became all the rage), and when she appeared on the White House steps with famed blind and deaf author Helen Keller to raise public awareness of the disabled, newsreel cameras whirred. "To Mrs. Coolidge, the country gives its heart," cooed the *Washington Post*, "which she has fairly won [through her] unfailing cheerfulness and charm."

Always stylishly dressed (for evening wear she favored brocaded gowns with long trains of gold lace), the steamboat inspector's daughter mixed easily with Washington's glitterati at White House receptions and state dinners, events her husband would just as soon have skipped. While greeting receiving lines, the president pulled his guests past him as he shook their hands, stifling any attempt at conversation. Seeing this, Grace would edge away from Calvin so she could chat with people. He would close the gap and she would move away again, and so it went.

First Lady Grace Coolidge wearing a brocaded gown in a portrait by Frank O. Salisbury.
(COURTESY OF THE VERMONT DIVISION FOR HISTORIC PRESERVATION.)

For all his lack of charm, President Coolidge was himself extremely popular. In a three-way race, he won the 1924 presidential election in a landslide by a margin even FDR, LBJ, and Ronald Reagan wouldn't match.

The victory was the crowning achievement of Coolidge's career, the culmination of twenty years of grueling campaigns—of all the endless whistle stops, stump speeches, and door-to-door vote hustling—and would have felt triumphant had it not come on the heels of a tragedy.

Calvin and Grace's younger son, sixteen-year-old Calvin Jr., tall, bright-eyed and witty, was his father's favorite. A few months before the election, the boy played tennis on the White House tennis court without his socks and developed an infected blister on his toe, which led to blood poisoning. Doctors were rushed in, but, without modern antibiotics, could do nothing. Within a few days Calvin Jr. slipped into a coma and died.

"In his suffering he was asking me to make him well," Coolidge wrote. "I could not." He had put on a brave face at his son's bedside, bringing the boy a pet rabbit, which made him smile. But when the end finally came, the president broke down, repeating through his tears, "I can't believe it has happened."

From then on, whenever he looked out his office window across the White House lawn, he saw Calvin Jr. there playing tennis. In his autobiography, Coolidge blamed the death on himself, writing that it would never have happened if he hadn't been president.

Plagued by guilt, haunted by his late son's memory, he lapsed into a deep depression. Her husband, Grace said, had lost his zest for living. He would faithfully perform his duties, hiding his anguish from the public, even smiling broadly when posing for pictures, but the world of politics no longer held any charm for him. When Calvin Jr. died, Coolidge wrote, "the power and glory of the presidency went with him."

A measure of solace came when Agent Starling saw a small boy with his face pressed against the fence outside the White House. The boy said he wanted to tell the president how sorry he was that his son died, so Starling took him to see Coolidge. Afterward, the president told Starling never to turn a boy away who wanted to see him.

After finishing his full four-year term, at the height of his popularity, when he could have coasted to re-election in another landslide, Coolidge gave up the presidency and withdrew from public life. "We draw our presidents from the people," he said. "I came from them. I wish to be one of them again."

After the 1928 election and the inauguration of the new president, Republican Herbert Hoover, Calvin and Grace moved back

Calvin and Grace with their two sons, John (left) and Calvin Jr. (right).
(COURTESY OF THE VERMONT HISTORICAL SOCIETY.)

into the Northampton duplex they had rented before he became governor of Massachusetts. If the ex-president expected to retreat into peaceful seclusion, he was sorely mistaken. Three thousand people greeted the Coolidges as their train pulled into Northampton, and many more lined the streets along their route home, cheering as the former first couple drove by. Almost from the moment they returned, an endless stream of gawkers paraded past their door, clogging the street with people and cars and annoying the former president no end. When he threw away a cigar butt, souvenir hunters scrambled for it like piranha, the winner proudly laying claim to a saliva-coated stub.

Coolidge's celebrity may have been annoying, but it was also profitable. His autobiography brought in tidy royalties and his syndicated newspaper column, "Thinking Things Over with Calvin Coolidge," earned him five figures a month, a whopping sum for the times. Along with the considerable sum he'd socked away during a lifetime of penny-pinching, it was enough to make him wealthy. After a year in the duplex, Calvin and Grace escaped the crowds by purchasing a nine-acre Northampton estate where Coolidge spent his remaining years.

Despite his wealth and the lush surroundings, contentment eluded him. Seven months after he left office, the stock market crashed and the country was soon mired in the Great Depression, a catastrophe some critics blamed on Coolidge. True or not, the criticism stung.

When the Democrats won back the White House and both houses of Congress in 1932, he became disillusioned. Just eight years after his landslide victory, his small-government conservatism was passé, replaced by President-elect Franklin Roosevelt's program of "bold, persistent experimentation." The world of politics where

THE PERFECT STICK

Coolidge had thrived for thirty years had passed him by. He told a visitor he felt ignored and forgotten.

Coolidge wouldn't live to see FDR inaugurated. At a ceremony at Amherst a few days after the election, Coolidge's classmates noticed how worn and tired he looked. Six weeks later, on January 4, 1933, he died at home of a heart attack. He was sixty years old.

He was laid to rest in the Plymouth Notch cemetery beneath a headstone inscribed with only his name and dates and the presidential seal. Grace, who survived him by twenty-four years and never remarried, lies beside him.

Al Smith, New York's Democratic ex-governor, called the late president "a shining public example of the simple and honest virtues which came down to him from his New England ancestors." H. L. Mencken, one of Coolidge's harshest critics while he was alive, wrote of him in death: "His failings are forgotten; the country remembers only the grateful fact that he left it alone."

Had there been room on his headstone for an epitaph, Silent Cal might have picked the nursery rhyme hanging on the wall of his Northampton home:

> A wise old owl lived in an oak
> The more he saw the less he spoke
> The less he spoke, the more he heard,
> Why can't we be like that old bird?

CHAPTER 5

SUZIE Q

Rocky Marciano

They say you can never tell the great ones until they hit the floor. It's how you make your fight with your eyes cut, your head spinning, your stomach aching and your nose bleeding that tells a champ.

—SPORTSWRITER JIM MURRAY

A right uppercut snapped Rocky Marciano's head back, and a left hook to his ear knocked him sideways and backward, nearly off his feet.

He caught himself on his back leg, staggered forward, and a moment later another left flush on the chin put him on his backside, the first time in a forty-three-fight pro career that Marciano had been knocked down.

It happened less than a minute into his title fight with world heavyweight champ Jersey Joe Walcott, and it set the stage for what followed. For twelve rounds, Marciano was, in the words of a ringside reporter, buffeted about like a cork on an angry sea. At thirty-eight (some claimed he was older), Walcott was past his prime yet still formidable, a clever boxer with power. The champ had trained hard and it showed. As Marciano lumbered forward, throwing roundhouse punches that whiffed past Walcott's chin, the champ stuck and moved, tagging the challenger with right leads, rocking him with left hooks when he dropped his guard.

Marciano kept coming, leaving Walcott's manager awestruck: "He got hit with some punches which would have knocked a building over and still he stayed in there."

Marciano scored in close with head and body shots, and won some of the middle rounds, but he never really hurt the champ, not badly anyway. In the eleventh, Marciano nearly went down again. By the end of the twelfth, he had a gash over his right eye and his left eye was nearly swollen shut. All Walcott had to do to keep his crown was last out the final three rounds.

Then it happened, the moment that turned an awkward brawler into a legend. With all due respect to another boxing legend, Jack Johnson, who in 1909 had landed a right cross that left Stanley Ketchel sprawled on the canvas next to his teeth, the right hook that flattened Jersey Joe in the thirteenth round on September 23, 1952, was hands down the hardest blow ever in a heavyweight championship fight.

No roundhouse punch, it started from just below Marciano's shoulder and traveled maybe eighteen inches. He threw it from his toes, packing every ounce of his 184 pounds into it and then some.

The champ never saw it coming. It slammed into his chin like a sledgehammer moving at lightning speed. "To ringsiders the sound of it was frightening," a reporter wrote. "It wasn't the smack of gloves against flesh. It was a crack." Backed against the ropes at the time, Walcott slumped to the canvas in a heap, face down, his left arm hung up on the bottom rope. The referee counted him out and then rolled him gently onto his back. Reviving him took several minutes. "I just don't remember anything," the dejected ex-champ said in the locker room. "I still don't know what hit me."

SUZIE Q

The heavyweight division had a new champion and his right fist had a name. Not the usual macho nickname the fight game came up with, like the Manassa Mauler (Jack Dempsey) or the Brown Bomber (Joe Louis), or like Marciano's nickname, the Brockton Blockbuster. This one was strictly tongue in cheek, a girlish name amusing to everyone but his opponents. As they could attest, you might outbox Marciano, beat him to the punch, cut him, stagger him, maybe even knock him down.

Right up to the moment Suzie Q kissed you good night.

He'd begun as a washed-up ballplayer digging ditches for a living. They said he was too old and slow for the ring. His trainer laughed

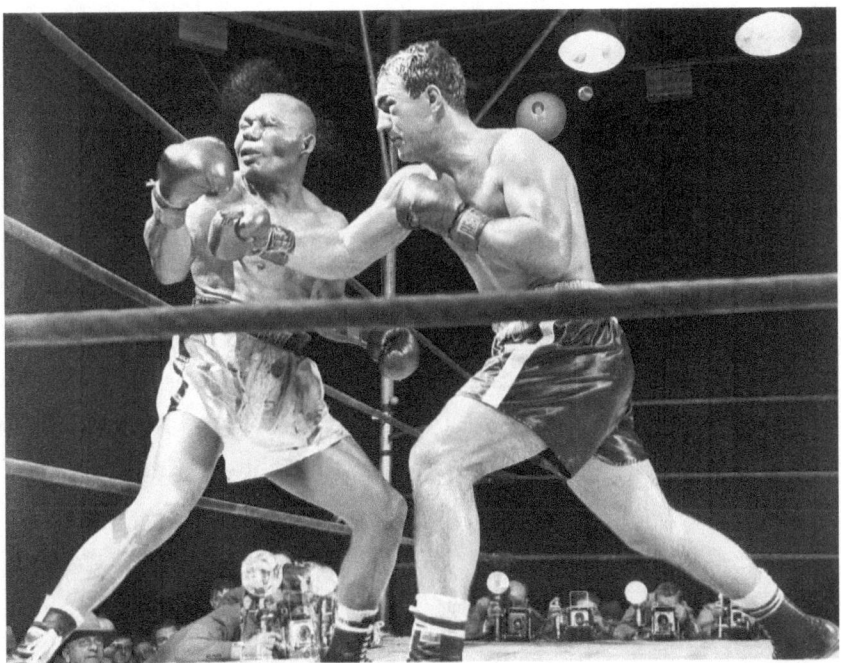

The punch that KO'd Jersey Joe Walcott. (GETTY IMAGES.)

the first time he saw him spar. Yet Rocco Francis Marchegiano, the son of poor Italian immigrants, became the greatest fighter of his era, maybe any era, the only heavyweight champion who never lost a professional fight.

He did it with grit, guts, maniacal drive, and a fighting style one opponent described as "Slam. Bam. Didn't give a damn."

Famed trainer Angelo Dundee sized Marciano up this way: "Two left feet. Stoop-shouldered and balding. But God could he punch."

Rocky Marciano was a class act. He didn't brag or talk trash or celebrate wildly in the ring after a victory. He didn't throw rabbit punches like Dempsey, or bite off part of an opponent's ear like Mike Tyson.

He occasionally landed low blows, but that was because of his free-swinging style, not because he was a dirty fighter. He had too much respect for his opponent and for the sport of boxing to intentionally hit below the belt.

Most of the time anyway. Not counting the night he tagged Henry Lester between the legs. Without using his fists.

The year was 1946. Marciano was a twenty-two-year-old amateur fighter home on leave from the army. He was chiefly a baseball player with dreams of becoming a big-league catcher. Boxing was just a sideline, a way to avoid KP duty. To keep from losing his amateur status, he took his purse for the Lester fight—all thirty bucks of it—under the table.

Lester was an ex–Golden Gloves champ. Marciano was a tub of guts. He'd spent his leave ballooning up on his mom's Italian cooking. His prefight meal was a plate of spaghetti and meatballs.

SUZIE Q

Most of Rocky's male relatives, not to mention half the population of his old neighborhood in Brockton, Massachusetts, showed up at Brockton's Ancient Order of Hibernians Hall to cheer him on. Family pride was at stake that night. Blubbery or not, he wasn't about to take a beating in front of a hometown crowd.

At the opening bell, to shouts of "Kill him, Rocky," Marciano charged forward. Halfway through round one he was sucking wind. By the second he was punched out. By the third, he stood at bay against the ropes, totally spent.

Lester moved in for the kill, and the next thing he knew he was flat on the canvas. Marciano had kneed him, to put it politely, in the groin.

Marciano fled the ring to a chorus of boos. "I'd never been so disgusted and embarrassed," he would recall. "I just sort of slipped home quietly that night." He vowed to never again be out of shape for a fight.

The next morning he made another vow. When he showed up at breakfast with a mouse under his eye, his mom got upset and insisted he give up boxing. "Sure, Ma," he said. "No more fights."

Pasqualena Marchegiano had wanted her eldest son to be the next Enrico Caruso. "But Ma," Rocky said, "I can't sing."

He also knew he didn't want to follow in his father's footsteps, operating a shoe-making machine in a sweatshop for forty years. "I'll never work in a shoe factory," the boy announced to his family. "I have to find a way out."

What Rocky wanted out of was a life of poverty in Brockton, a working-class city of sixty thousand, twenty miles from Boston and light years from the glamor of big-time sports.

GLORY, GRIT AND GREATNESS

Scraping by during the Great Depression, Rocky, his Italian-immigrant parents, and his four siblings shared a few cramped rooms upstairs in his grandparents' house, without hot running water, central heat, or a bathtub. Rocky slept on a cot in the parlor. The family bathed every other Saturday from a washtub in the kitchen. There was plenty of food on the table, but the kids wore hand-me-down clothes and there was little room in the family budget for sports equipment. Broken baseball bats were screwed back together; worn out baseballs were wrapped in carpenter's tape.

Meanwhile, just across the tracks, lay the opulence Rocky dreamed of. In winter, he and his friends Izzy and Eugene, coal shovels slung over their shoulders, would traipse in their hip boots to the posh side of town to shovel snow off the driveways of Brockton's upper crust. Like kids at a candy-store window, the boys gaped at the sprawling brick homes and the shiny Packards and Cadillacs in the driveways, their whitewalls and grillwork gleaming as brightly as the fresh-fallen snow. Rocky came away dazzled.

In his own neighborhood he was becoming known as a kid not to mess with. Generally quiet and mild-mannered, "a perfect gentleman," said his fifth-grade teacher, when riled he became ferocious, "like an animal," a friend said. When he was thirteen he decked a bully who called Eugene a dumb Guinea, sending the kid to the hospital for stitches. The punch that did it was a right hook.

Hardscrabble immigrant grit ran in the family. While Rocky's dad, Pierino Marchegiano, was fighting in the trenches during World War I, a piece of shrapnel slammed into his jaw, knocking out three teeth. "Didn't bother me though," he told reporters thirty years later. "I spit them out and kept going."

Pasqualena was a gregarious fireplug of a woman, strong-willed and strong-limbed (Rocky always claimed he got his brute strength from his mom), who outweighed Pierino by fifty pounds and wore the pants in the family. Her father, Luigi, was a retired blacksmith. On Saturday nights, he and his paisans, plied with red wine, settled disputes over their card games by charging each other from across the room and banging their skulls together like bighorn sheep. The crack it made echoed to the top of the stairs where young Rocky watched, wide-eyed.

While Pierino labored at the shoe factory, Pasqualena kept house and cooked lovingly prepared Italian dishes—spaghetti and meatballs (her specialty), *spezzatino* (Italian beef stew), and salads made with tomatoes and zucchini from her father's garden. "We never had much in those days," she recalled. "But there was always a lot of love in our house and we knew that someday it would be better."

Rocky's chance for something better lay in athletics. He was a powerhouse on his high school football team, playing sixty minutes a game at center and linebacker ("a rough, tough, powerful kid of about a hundred and fifty-five pounds who never got tired and never got hurt," the coach said), but he was too small and slow for the NFL.

Mostly, he pinned his hopes on baseball. Blessed with a strong bat and an arm that was accurate if not overpowering, the aspiring young catcher lived and breathed the game, playing every chance he got, for his high school, for his church, on an American Legion club, whatever team would have him. After games, he practiced his throwing and hitting until it was too dark to see the ball. For upper body strength, he did chin-ups from a tree branch in his backyard.

To help support his family, he put his ambitions on hold and dropped out of high school to work at odd jobs: delivery truck driver, short-order cook, dishwasher, ditch digger. He even worked in his father's shoe factory for a while. With the coming of World War II, Rocky was drafted into the U.S. Army and served in an engineer battalion, ferrying supplies to the troops fighting in France.

Honorably discharged, he lined up a tryout with a boxing trainer. The fellow was left shaking his head. The kid was a lummox, hopelessly clumsy. His arms were too short. His legs were too thick. He was too small for a heavyweight and too big for a light heavyweight. And, at twenty-three, he was too old to start a ring career.

"Forget boxing, boy," the trainer said. "You'll get killed."

It is now September 17, 1954. Thirty-four thousand fight fans in Yankee Stadium are screaming for blood. And getting it.

As thirty-one-year-old heavyweight champ Rocky Marciano sits motionless on the wood stool in his corner with his head tilted back, his cut man Freddie Brown is working frantically to stanch the blood pouring from the fighter's nose.

In all his years in the ring, Brown has never seen a cut like this one. It's a ghastly wound, the result of challenger Ezzard Charles grazing the champ's face with an uppercut in the sixth round of their title fight. Marciano's left nostril is sliced clean through and the blood is flowing so freely the nose reminds Brown of a faucet. Across the ring, Charles's corner is ecstatic. Maybe the fight will be stopped.

Seconds tick by. Brown has all of one minute to fix a wound that will need stitches and take weeks to heal. The ring doctor climbs through the ropes, eyes the cut and gives the okay for Marciano to continue. Brown applies a powerful coagulant and puts a patch on Marciano's nose. It looks silly but it works. The bleeding slows, at least for now.

In round seven, Charles, a proud ex-champion fighting for his life, throws every punch in the book at Marciano's nose. The champ fights back hard and staggers Charles as the round ends. Marciano is way ahead on points, but the patch has been knocked off and the cut is open again. The referee tells Marciano's corner he'll let the fight go one more round, and then he'll have to stop it. Marciano's

Marciano with his father, Pierino, and his mother, Pasqualena.
(BOTH PHOTOS FROM GETTY IMAGES.)

manager screams at his fighter, "Go after him. You gotta knock him out."

Marciano isn't about to panic. He's used to fighting back from the brink of disaster. In the opening rounds of his first fight with Charles three months earlier, the challenger beat him to the punch

 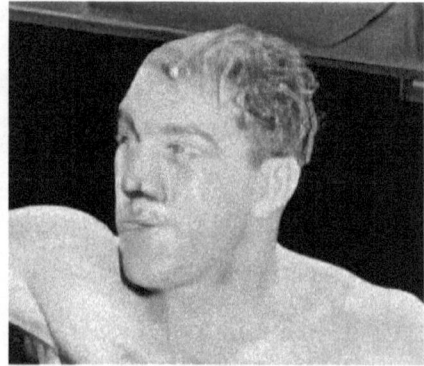

Left: Marciano vs. Ezzard Charles, September 17, 1954. (GETTY IMAGES.)
Right: The cut on Marciano's nose.

time and again, cutting him and bloodying his nose, before Marciano rallied to win a unanimous decision. The night he won the title from Jersey Joe, Marciano fought three rounds with ointment burning his eyes, barely able to see.

Now, with his title on the line, he knows what he needs to do and proceeds to do it. Seconds into round eight, Charles opens a gash over Marciano's eye. Rocky doesn't even blink. A minute later, bleeding from his eye and his nose, he floors Charles with a looping right to the jaw—Suzie Q. Charles struggles to his feet. Marciano bores in and connects with three combinations, six straight blows to the head. This time Charles doesn't get up until the referee has counted him out. Marciano has survived the cut and saved his crown.

In the dressing room, his nose heavily bandaged, the champ tells reporters, "I like my title too much to lose it on account of a little blood."

Seven years earlier, in the spring of 1947, twenty-three-year-old Rocky had piled into a car with some other baseball hopefuls from

Brockton, and headed for Fayetteville, North Carolina. He had landed a tryout with a team in the Chicago Cubs farm system, the Class B Fayetteville Cubs.

As quickly as Rocky's baseball dreams caught fire, they flickered out. He lasted just three weeks before the club cut him. He tried out for two other minor league clubs in the South that spring with the same result. Gifted enough to star on sandlot teams back home, he simply didn't have big-league talent.

On the long train ride back to Brockton, he sat staring out the window. All those years playing ball in the summer heat with worn-out equipment, the countless hours practicing his hitting and throwing, the thousands of chin-ups, had been for naught. He'd taken his shot and failed. What would he do now? He was sharing the ride with a friend who'd also been cut. After a while, Rocky turned and said, "The heck with it. I'm through with baseball. I'm gonna get some fights."

From then on, winning the heavyweight title was his obsession. No fighter—no athlete—trained harder. He would put in a day's work digging ditches for the Brockton Gas Company, then do several miles of roadwork and spend a couple hours punching a seabag full of sawdust.

Later, when boxing became his full-time job, he trained as many as twelve hours a day, lifting weights, skipping rope, hopping up flights of stairs on one leg, shadow boxing in shoulder deep water, bouncing a basketball for coordination, even training his eyes by lying face up on his bed and tracking a pendulum he'd rigged overhead.

Too slow and awkward to outbox top heavyweights, he outworked them. Other fighters trained for weeks before a fight. Rocky trained for months. There were those who claimed he enjoyed his grueling regimen, and perhaps he did. He was that driven. The hardest part was giving up his mom's cooking. Spaghetti was no longer on the menu. Dinner consisted of a rare steak, chewed but not swallowed, and a salad.

Between fights he was known to spar over two hundred rounds, far more than most fighters. "Rocky was brutal even in sparring," his kid brother and sometime sparring partner, Sonny, remembered. "He just didn't know any other way." One time, Sonny made the mistake of tagging Rocky with an unusually hard right. He spit out his bloody mouthpiece, banged his gloves together, and began beating his brother to a pulp. "I knew he wanted to murder me," Sonny said, "but there was nothing I could do." Sonny screamed, "Rock. Rock. Take it easy. I'm your brother." That snapped Rocky out of it and he backed off.

Saturday evenings, after his workout, he sometimes tagged along with friends to a social club, where one night he spotted a tall brunette on the dance floor. Barbara Cousins, a telephone operator, the daughter of a Brockton cop, was a fun-loving eighteen-year-old. Rocky was a twenty-four-year-old wallflower who stepped on her feet the first time they danced. He was also a dark-eyed hunk. Mutually smitten, before long they were engaged. His all-consuming quest for the title, and later his philandering, would take a heavy toll on their marriage.

When Rocky decided to turn pro, he had to go all the way to New York City to find a manager willing to take on an unschooled pug in his mid-twenties. That manager was, of all things, a former professional ballroom dancer. Rumor had it that when he won a dance contest, he would make off with the prize money while his partner was busy taking a bow. You trusted Al Weill at your peril.

White-haired and portly, flashing a diamond pinkie ring and an ever-present stogie, forever throwing his weight around in a bullhorn voice with a Brooklyn accent, Weill came across like a loudmouth mobster, which wasn't far from the truth. His mob connections made him a powerful force in the fight game.

Life magazine called him "the most cordially hated manager in boxing." He was also among the most effective. Three of his fighters had already become champions. He was a master at orchestrating a fighter's career, moving him up the ranks by handpicking his opponents. When a fighter was ready, Weill had the clout to get him the big-time fights he needed to earn a title shot.

Weill ruled his fighters with an iron fist. He told them where to live, what to eat, what to wear, even when to get married. And he didn't come cheap. Marciano had to sign over half of everything he made, not just during his boxing career but in retirement as well. Eventually, he grew to detest Weill, while admitting that without him he would never have become champion.

Marciano's trainer, the man who taught him how to fight, came straight out of central casting.

Israel "Charley" Goldman was the model for the gnarled old trainer portrayed by Burgess Meredith in the *Rocky* movies. An ex-flyweight, all of five feet tall and 105 pounds, Charley fought his first official match in a Brooklyn saloon when he was sixteen, twenty years before Marciano was born, at a time when boxing was illegal in New York. The fight went forty-two rounds and was finally called when the cops raided the joint.

Goldman fought 137 bouts and lost only 6. And that was just according to the record book. The true number was more like 400, not counting the times in the fourth grade he ditched school to fight for a few nickels. He fought for pocket money in the back rooms of taverns, in dance halls, and in countless boxing clubs, sometimes twice a day.

Top Left: Mr. and Mrs. Rocky Marciano. (STANLEY BAUMAN COLLECTION/STONEHILL COLLEGE.) *Top Right*: Al Weill. (ALAMY.) *Bottom*: Charley Goldman with Marciano. (GETTY IMAGES.)

By the time he hung up his gloves, he had cauliflower ears, a flattened nose, two fistfuls of busted knuckles, and an encyclopedic knowledge of how to prevail in the ring. He became the consummate tutor of boxing skills. Ex-pugs-turned-trainer aren't usually considered geniuses, but that label stuck to Charley. He was known in the fight game as the "Ol' Perfesser." Rocky Marciano would be his masterwork, "a block of marble," a sportswriter wrote, that Goldman "sculpted . . . into the Pietà."

Goldman never forgot the first time he saw Marciano spar:

I'll eat my derby hat if I ever saw anyone cruder than Rocky. He was so awkward that we stood there and laughed. He didn't stand right. He didn't throw a punch right. He didn't block right. He didn't do anything right. Then he [landed] a roundhouse right which nearly put a hole in the guy's head, and I told Weill that maybe I could do something with him.

"I've got enough broken-down fighters already," Weill snorted. He refused to pay Rocky's expenses while Goldman trained him.

Unemployed at the time, Rocky bummed his expenses off his friend and cornerman Allie Columbo. The two of them and another fighter shared a room at a YMCA in New York for $1.70 a night. Every week or two, Rocky returned home for a match on fight night at the Rhode Island Auditorium in Providence, thirty-five miles from Brockton. All told, more than half his pro fights would be in Providence. He started out fighting on the undercard, four rounds for $40, and worked his way up to the main event, ten rounds for $250. Afterward, he'd hitch a ride back to New York to resume training.

GLORY, GRIT AND GREATNESS

Under Goldman's tutelage, Rocky's signature style began to take shape, that swarming, smothering, relentless attack, the constant slugging from every angle that ground down opponents and set them up for the kill. He turned his fights from contests of skill into wars of attrition, bloody tests of stamina and heart. By that measure he was unbeatable. No one was tougher.

And no one hit harder. It was said that his right fist "registered nine on the Richter scale," but it took Goldman to harness the God-given power. He tied a string around Marciano's ankles to shorten his stance and improve his balance and leverage, and tied Rocky's right hand behind his back so he would learn to use what would become a crushing left.

Obeying his favorite mantra—"If you got a short guy make him shorter"—Goldman turned Rocky's lack of height (he was five-foot-eleven) to his advantage. He taught his prize pupil to fight from a deep crouch, bobbing and weaving, then lashing out like a coiled snake.

For those who said Rocky looked clumsy, his trainer had the perfect retort: "The guy on the ground don't look too good either."

Knockouts started piling up. By the end of 1949, Rocky's record was 24–0, with twenty-two knockouts, nine in the first round. Even though most of his opponents were journeymen or outright stiffs ("slightly animated punching bags," one journalist called them), it was an impressive streak.

The press and public took note, and so did Al Weill. Before long, he sewed his new prospect up with a written contract. He also had him change his name. The ring announcer in Providence kept butchering it. So Rocco Marchegiano became Rocky Marciano.

Then on December 30, 1949, he nearly killed a man.

SUZIE Q

A ringside snapshot told the story. It showed Carmine Vingo out cold on the canvas, his face a mass of blood, his jaw sagging, his eyes swollen shut, while a cornerman knelt over him cradling his head.

Like Marciano, Vingo was a promising young slugger, bigger than Marciano and nearly as powerful. For six rounds they'd stood toe-to-toe beating each other bloody, until Marciano knocked him out.

Vingo lapsed into a coma. Marciano spent an agonizing vigil at the hospital with Vingo's family, and paid a large portion of Vingo's medical bills. Eventually, Vingo recovered and he and Marciano became friends. Vingo remembered Marciano as "one of the nicest guys you'd ever want to talk to."

Through it all, Marciano remained as driven as ever. "Don't worry. I'm going for the title," he assured Charley Goldman, "and nothing's going to stand in my way."

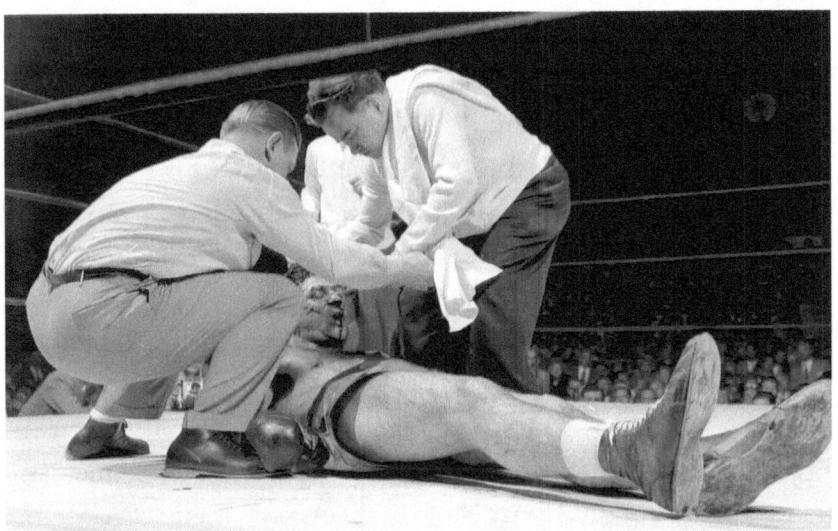

Marciano victim Carmine Vingo. (GETTY IMAGES.)

And nothing did. Not the Vingo fight. Not an arthritic right elbow that made it hard to straighten his arm. Not a broken right hand he suffered throwing a punch. (He went right on training with a cast on.) Not even a ruptured disc, a condition that occurred early in his career, and that he kept hidden from Weill for fear he would drop him. A doctor told Marciano he needed back surgery and would have to quit boxing for at least a year, maybe permanently.

He ignored the advice, and the pain, and kept fighting. Visits to a chiropractor and having his legs pulled twice a day helped, although his back would bother him off and on until he retired. The night he fought Pete Louthis in August 1949, his back hurt so much he could barely climb through the ropes. He knocked Louthis out in the third round.

In Marciano's first big fight, he eked out a controversial split decision against top contender Roland La Starza. Burly Rex Lane, "the next Jack Dempsey," fell in six—"like an elephant collapsing from a rifle shot"—a victim of Suzie Q.

After that, the last big hurdle on the way to a title shot was one of Marciano's idols, the great Brown Bomber himself.

Like the rest of the country, fourteen-year-old Rocky had his ear glued to the radio the night in 1938 Joe Louis put his heavyweight title on the line against Germany's Max Schmeling. Louis had a score to settle (two years earlier Schmeling had knocked him out), but as far as America was concerned, he was squaring off against Adolf Hitler and Naziism. When Louis knocked Schmeling out in the first round, the country cheered.

Louis emerged from World War II—during which he won the Legion of Merit for donating his purses to needy servicemen and doing morale-building tours while in the Army—as one of America's most admired figures. It didn't matter that Marciano was the Great White Hope. When they fought in Madison Square Garden in October 1951, the mostly white crowd was rooting for Louis.

SUZIE Q

By then, the man regarded by many as the greatest fighter ever was a sad shell of himself, an aging ex-champ making a comeback because he owed money to the IRS. The only weapon left was his jab, and early on he stung Marciano with it, bloodying his nose and cutting him under the eye. It wasn't enough. While the crowd cringed, Marciano battered his weary idol into submission, finally landing a right that knocked him clear through the ropes in Round 8. The referee didn't bother finishing the count.

"I'm glad I won," Marciano told reporters, "but I'm sorry I had to do it to him." Outside the ring, Marciano had regrets. (One report had him weeping when he went to Louis's dressing room to console him.) Inside the ring, his idol had been just another obstacle on the way to the title. "I want to kayo Louis because it will move me closer to the championship of the world," he'd told reporters before the fight. "To me, being champ means everything."

Marciano was now earning five-figure purses, and the first thing he did with his newfound wealth was see to it that his dad quit his job. After laboring for forty years on an assembly line, Pierino Marchegiano retired from the shoe factory thanks to the generosity of his eldest son.

It was the biggest event in Brockton's history. Schools and businesses closed in honor of the occasion. Upwards of sixty thousand people lined the parade route, more than had greeted Presidents Franklin Roosevelt and Harry Truman when they'd visited the city—more, in fact, than the city's entire population. Packed along the sidewalk downtown, with the overflow spilling into the street, dressed in their Sunday best on that muggy Thursday afternoon in the fall of 1952, they'd turned out to welcome home the pride of Brockton, the newly crowned heavyweight champion of the world.

GLORY, GRIT AND GREATNESS

Cops lined shoulder-to-shoulder held back the throngs, as the motorcade rolled by under showers of confetti. That night at a testimonial dinner a group of local businessmen, who'd made a killing betting on the champ, rewarded him with a new Cadillac. The governor of Massachusetts chipped in the license plate. It read—what else?—"K.O."

The kid who'd worn hand-me-down clothes and shoveled snow for pocket change had become a megastar. For the rest of his short life, Rocky Marciano would ride a wave of stardom that would bring out the best in him, and the worst.

Ducking no one, he defended his title six times. Jersey Joe Walcott fell in the first round of their rematch. Roland LaStarza in eleven. Pudgy, overmatched British champ Don Cockell, "The Waist of Time," went in nine.

Some other white heavyweight champs, Jack Dempsey included, had refused to fight African Americans. Marciano not only fought them, he won their admiration. Joe Louis called him "one of the finest fellows I know" and "a personal friend." To Jersey Joe, Marciano was "not only a great champ, but a great American."

His color blindness extended to his fans. *Los Angeles Times* sportswriter Jim Murray told of the time four black men drove 250 miles from Philadelphia to the Catskills to watch Marciano spar, only to find it was the champ's day off:

> I ran to Rocky. "Those guys drove all the way up from Philly," I told him.... He put [his] book down and hurried over to them. ... The fellas all took their hats off to shake hands with the champ.... Rocky chatted with them for half an hour. *They* broke it off."

That was the Rocky everyone knew and loved, the kindhearted celebrity with the common touch, who signed autographs till his hand hurt and played Santa Claus at the City of Brockton's annual Christmas party, who made countless charity appearances, visited sick kids in an inner-city hospital, and gave a $10,000 championship belt to Ezzard Charles when he was penniless and suffering from Lou Gehrig's disease.

There was a seamier side few people saw and no one fully understood. "He was a strange one," a friend once said. "Nobody ever knew Rocky except Rocky, and that was exactly the way he wanted it." His final bout was vintage Marciano. The Rock at his gritty best. Legendary light-heavyweight champ Archie Moore beat him bloody and knocked him down for only the second time in his career. As he had so many times before, Marciano shook off the beating and kept swinging, dropping Moore three times before taking him out in nine. The crusher was a textbook left hook to the chin, a short, cat-quick stroke, just like Charley Goldman had taught him.

As soon as the ref quit counting, Marciano rushed over to check on Moore and help get him to his feet, and then praised him to a television reporter: "Archie's a very tough, willing, determined fighter. I didn't think he would stand up like he did.... Congratulations to him." That too, was Rocky at his best.

Seven months later in the spring of 1956, his record 49–0, he stunned the boxing world and retired at age thirty-two, becoming the only undefeated heavyweight champ in history.

Publicly he gave the usual excuse: he wanted to spend more time with his family. In truth, he was burned out. Training had become a grind. He didn't want to become another punch-drunk fighter who'd hung on too long. He wanted to go out on top, with his marbles and dignity intact. But mostly, he was fed up with Al Weill's bullying, chiseling ways. When Marciano heard Weill had skimmed ten grand from one of the champ's title fights, it was the last straw.

GLORY, GRIT AND GREATNESS

Other great champs—Joe Louis, Muhammad Ali, Larry Holmes—embarrassed themselves by making comebacks. Marciano stayed retired, turning down million-dollar offers to fight again. He came close to returning to the ring in 1959 when Ingemar ("The Hammer of Thor") Johansson from Sweden won the title. Convinced he could beat him, Marciano went into training, until he realized he was too old and flabby and changed his mind.

The press portrayed the ex-champ as a devoted family man retiring to a life of domestic bliss with his wife and three-year-old daughter. The truth would have made better copy. More lurid anyway. Marciano spent the better part of his retirement on the move, crisscrossing the country in a mad rush to cash in on his fame, hanging out with gangsters and movie stars, leading the life of a swinging bachelor, while his wife and daughter stayed behind at home.

"Rocky chased a buck," a sportswriter quipped, "the way he chased Ezzard Charles." Al Weill put it more bluntly: Giving Marciano money "was like feeding blood to a tiger. He just couldn't get enough."

There were myriad ways a retired sports icon could cash in, and Rocky tried them all. He had his own talk show, *The Main Event*, for a while, did color commentary at ringside for ABC, endorsed products on television, scored the occasional bit part in a TV western, and invested in countless businesses—a sausage company in Ohio, a chain of Italian restaurants in California, a bowling alley in Florida, and so on—some of which made money, some of which didn't. No matter. He could always recoup his losses with personal appearances. Seemingly everyone wanted to bask in the aura of history's only undefeated heavyweight champ, and those who could afford it were willing to pay handsomely for the privilege—between $2,500 and $5,000, plus expenses, for a dinner speech, at a time when five grand bought a new Cadillac.

Money flowed in, a river of cash. Whenever possible, Marciano demanded cash. His accountant Frank Saccone once saw him being handed a cashier's check for $5,000 to cover his speaking fee. "I don't take checks," Marciano said shaking his head. Saccone pulled him aside: "Look, Rock. Cashier's checks are guaranteed. They're as good as money. I'll cash it for you. There's nothing to worry about." Marciano told him to butt out. "Can you bring me twenty-five hundred dollars?" he asked his host. The man left and returned with the money. Marciano pocketed it and returned the check.

Mama Pasqualena hadn't raised a fool. With cash, her son could hide his income from the grubby mitts of Al Weill, who was still entitled to half his income, and the even grubbier mitts of the IRS. Federal tax rates in the 1950s ran as high as ninety percent. Between them, Weill and the IRS had siphoned off a fortune from the money Marciano had earned getting pummeled in the ring. He figured enough was enough.

He also shied away from bank accounts. They left a paper trail for the IRS. Instead, he rat-holed his cash in whatever hiding place was at hand: in a curtain rod, in the hole above a light fixture, or wrapped in plastic and taped under the lid of a toilet tank.

Tales abound of him carting around grocery bags and suitcases full of cash. He would show up in Brockton and dump five or six grand in cash on his mom's kitchen table. "What's this for?" she asked. "Spending money, Ma. Spending money."

For someone who went after money like a tiger, he could be maddeningly careless with it. Once as he was leaving a theater, he accidentally left a paper bag behind on his seat. His daughter spotted it and picked it up. It contained $40,000.

He was flush with cash partly because the whole time he was traveling around making money, he hardly spent a dime. Wherever he went, there were well-heeled admirers glad to pay his way, and he

Post retirement: Marciano (*top*) with his wife and daughter. (GETTY IMAGES.)
Bottom: with Hollywood stars Zsa Zsa Gabor (LEFT; ASSOCIATED PRESS.)
and Mamie van Doren (RIGHT; GETTY IMAGES.)

gladly let them. Hotel suites, flights on private planes, limos, meals at swanky restaurants—the whole red-carpet treatment came free of charge. He "could leave town with thirty cents in his pocket," his chauffeur remembered, "and tour the country and he'd come back with the same thirty cents." Meanwhile, the factory worker's son always made sure that whatever rich admirer he happened to be out with left a hefty tip for the waiter and doorman.

More than a few of those admirers were mobsters—like Frankie Carbo, the reputed ex-hitman, and Peter DiGravio, the alleged consiglieri of the Cleveland mob. And Vito Genovese, "the boss of all bosses."

"Wherever we went there were mob guys," said Richie Paterniti, Marciano's longtime friend. "They loved him because he represented what mob guys really want to be, the toughest guy in the world, right? A macho guy. They all had respect for him. They all wanted to be with him. They kissed his ass. Every mob guy. He was an Italian, and he beat up every guy he faced. He exuded power, an air of authority. That's why they wanted to bask in his sunshine."

One night, accountant Saccone found himself sitting with Marciano at a table full of fawning wiseguys. "They couldn't do enough for him. They'd say to him, 'I got a beautiful tailor. Let me take you down there and get you some suits.' They'd buy him six suits, three-dozen shirts. He loved it and they loved it."

Although Marciano hobnobbed with gangsters and loaned them money (DiGravio borrowed $100,000 from him, then got whacked before he could pay it back), he didn't involve himself in their criminal activity, during his ring career or after it. When a syndicate bigwig showed up before the Cockell fight and promised to make Marciano rich if he took a dive, Rocky threw the man out of his hotel room. Surrounded by mobsters, crooked managers, and fixed fights, Marciano kept his integrity intact. He "stood out," mused boxing columnist Jimmy Cannon, "like a rose in a garbage dump."

Not that Marciano was a saint. His retirement years were sullied by temper flare-ups and womanizing. When a telephone operator refused to return his dime for a call that didn't go through, he ripped out every receiver in an entire bank of pay phones. That incident didn't make the papers, but when he slugged a writer and got tagged with a $5,000 personal-injury judgment, it brought some bad publicity, along with a stern lecture from the judge about the importance of restraint.

He also became what one friend called "the heavyweight champion of girls." As brother Sonny recalled: "The girls used to fall all over him. We'd go into Toots Shor's and they'd be with Dean Martin and everybody wanted to sit with Rocky. And Frank Sinatra—they all wanted to be at Rocky's table. And the young girls who tagged along would be in awe of Rocky."

He didn't bother with love affairs, just quickies, an endless string of one-night stands with groupies and call girls. Prearranged trysts became a regular part of his city-hopping routine. A celebrity gig, a midnight romp, and he'd be off to the next town where yet another buxom blonde awaited.

Sadly, he'd long since lost interest in the woman he'd fallen for and married. Neglected by her husband, struggling with depression born of loneliness and five miscarriages, overweight and drinking heavily, Barbara, the pert young brunette, had grown old before her time.

Even before Rocky quit the ring, his long absences at his training camp in the Catskill Mountains had put a strain on their marriage. After he retired and became a skirt chaser, Barbara threatened divorce, but he talked her out of it.

While she sought solace in the bottle, he assuaged his guilt by keeping her and their daughter Mary Ann living in style in a beach house in Fort Lauderdale, Florida. He lavished furs and jewelry on Barbara, bought Mary Ann a new Firebird before she was old enough to drive, and took the family on lavish vacations between his road trips.

A year before his death, he and Barbara adopted a baby boy, Rocco Kevin, who looked a lot like Rocky and, it was whispered, was probably his illegitimate son. The baby brought a ray of sunshine back into the marriage.

Home to Fort Lauderdale to be with his wife and children was where Marciano was bound that Sunday afternoon, August 31, 1969, for what promised to be a joyous homecoming. There were birthdays to celebrate—Barbara had just turned forty and Rocky's birthday was the next day—but his wife had also planned a special surprise. As Rocky came through the front door, he would see Rocco Kevin walk for the first time, as he toddled forward to bring his dad his birthday gifts. The presents were wrapped, the birthday cake was decked with candles, and the guests had already arrived when Rocky called from Chicago to say he had a quick stop to make. A friend was opening an insurance business in Des Moines and had asked Rocky to make an appearance. "Hold everything," he told Barbara. He wouldn't be long and would fly home immediately afterward.

Late that evening, when Mary Ann heard the doorbell ring and her mom scream, she guessed the awful truth, even before hearing it from the police officer at the front door. The Cessna her father had been riding in, flown in bad weather by an inexperienced pilot, had crashed in an Iowa cornfield, killing all aboard. One day short of his forty-sixth birthday, Rocky Marciano, the invincible iron man of boxing, was dead.

"He was a wonderful father. . . . All my memories of him are good," Mary Ann would say decades later. "He'd pick me up and spin me on his knee [and] walk into little Rocco's room and pick him up and say, 'My son. My son.' His death shattered the family. When he died, a piece of my mother died."

A heavy smoker who fell victim to lung cancer, Barbara would follow her husband to the grave just five years later. During her final delirious hours she cried out, "Rocky, I'm not ready to go yet." Having spent much of their married life apart, today they lie beside each other in a Fort Lauderdale cemetery.

He left no will, and whatever money he'd hidden away was never found. One thing he did leave behind was a note he'd scribbled in a moment of reflection. It read: "Live fast, die hard."

In death, the poor immigrants' son who became a boxing icon received many a tribute. Congressmen eulogized him on the floor of the House of Representatives. Pulitzer-Prize–winning sportswriter Red Smith called him the "toughest, strongest, most completely dedicated fighter who ever wore gloves." Jim Murray, another Pulitzer Prize winner, wrote, "Rocky retired undefeated. He died the same way. 'Champ' will never have quite the same meaning for me."

But the most glowing tribute was unspoken. After arriving at the funeral service and mingling quietly with the crowd, a balding hulk of a man in a gray suit broke away and walked toward the altar.

What seemed like a lifetime ago, in front of seventeen thousand screaming fans in Madison Square Garden, he'd taken one of the worst beatings Marciano had ever dished out.

This time, in the hush of a Florida church, Joe Louis, the Brown Bomber, leaned down and kissed the late champ's casket.

CHAPTER 6

HEROES OF THE LAKES

Oliver Hazard Perry and Thomas Macdonough

"Clear for action!"

Aboard the American flagship, the *Lawrence*, the order sent the hundred-man crew into a flurry of activity.

Gunners hefted thirty-two-pound cannon balls onto wooden racks. The surgeon donned his butcher's apron and laid out his probes and bone saws. Hatches were secured and stands of cutlasses set up, along with barrels of water for putting out fires. Sand was sprinkled over the deck so that the living wouldn't slip on the blood and entrails of the slain.

With everything in readiness, an uneasy calm settled over the ship as it closed on the British fleet. In a grim ritual, the crew downed a nerve-steadying ration of grog. Some men scrawled wills on pieces of paper spread over barrelheads. Others handed shipmates farewell letters to loved ones. The *Lawrence*'s young captain went below and fished through his sea chest for a packet of letters from his wife Betsey. Finding them, he hesitated, then tore them up and had them thrown overboard, lest they fall into the hands of the British.

The enemy fleet was still a mile away when a round of solid shot tore through the *Lawrence*'s wooden hull, killing one of the crew. "Steady boys," the captain said. He ordered his gunners to hold their fire. The *Lawrence*'s guns were more powerful than those of the British, but of shorter range. Returning fire at such a distance would be a waste of ammunition.

He was a cool one, this twenty-eight-year-old Rhode Islander, brave to the point of recklessness. Handsome and charismatic, with a name as dashing as his looks, he seemed perfectly cast in the role of a great naval hero. Yet Oliver Hazard Perry had never been in a sea battle and had never commanded anything larger than a schooner, let alone anything like the motley flotilla he was now leading into battle. How he would perform was anyone's guess.

It was September 10, 1813, on the western waters of Lake Erie south of Detroit, Michigan. Thirty years after America had won its independence from Great Britain in the Revolutionary War, America and Britain were at war again, a conflict known to history as the War of 1812.

Like Perry himself, America was in a fight for its life. Having seized Detroit, the British were poised for a push farther south into the Ohio River Valley, America's northwest frontier. First, they needed control of the region's only good supply route, Lake Erie. If Perry could drive the British off the lake, the threat would collapse.

After struggling for months to build a fleet in the middle of the wilderness, he was sailing into battle with patchwork ships, green crews, and a treacherous second-in-command.

Despite it all, Perry was determined to attack. Moving slowly in the light breeze, the *Lawrence* plodded forward, as dozens of long-range guns zeroed in on her.

Master Commandant Oliver Hazard Perry.
(LIBRARY OF CONGRESS.)

Five hundred miles northeast of Lake Erie, dagger-shaped Lake Champlain knifes its way south from the Canadian border into the heart of upstate New York.

Known today for bass fishing and romantic getaways, in fiercer times the lake region was a killing ground. French explorer Samuel de Champlain discovered the lake in 1609 while paddling with an Algonquin war party to raid a Mohawk village, and for the next two centuries, French, British, and American soldiers, along with Native American braves, crossed the region time and again to wage war on each other, until the French were driven out of Canada, the Indians were driven west, and America won its independence.

During the War of 1812, the British, the conquerors of Canada, decided to use the age-old invasion route one last time. In the summer of 1814, they sent eleven thousand crack troops, His Majesty's finest, marching south along the lake, threatening to split New York state in half and reclaim New England for the British Empire.

On land, the juggernaut looked unstoppable. The Americans' only hope was to beat the Royal Navy on the lake, which would cut the British army's supply line and force a retreat.

In command of the American flotilla at the Battle of Lake Champlain, fought a year and a day after the Battle of Lake Erie, was a deeply religious thirty-year-old from Delaware, a Christian gentleman with a warrior's heart, as pious as a saint and as sly as a fox.

Pious in church, in action Thomas Macdonough was a swashbuckler.

In 1804, the twenty-year-old midshipman had been among a squad of night raiders assigned to sneak aboard an American frigate captured by Barbary pirates and burn her as she lay anchored in Tripoli Harbor, off the coast of Libya. After a slashing, hand-to-hand skirmish in which twenty pirates were slain, Macdonough stayed behind to set the frigate ablaze, escaping just ahead of the flames. The renowned British Admiral Horatio Nelson reportedly called the raid "the most bold and daring act of the age." Macdonough was lauded for his heroism. His star had begun its rise.

Six months later, during another shipboard skirmish, he wrestled a pistol away from a huge Turk and killed him with it. Macdonough's "zeal, courage and readiness" were cited in the official report of the action. Ashore in Sicily, three thugs made the mistake of trying to rob him at dagger point. Macdonough stabbed two and chased the third up a building, scaling it with his cutlass clamped in his teeth. The robber became so terrified he jumped off the roof to his death.

When not terrorizing his enemies, Macdonough was a model of Christian decorum, "high-minded, honorable and religious," said a shipmate. Macdonough didn't drink, gamble, or swear. His strongest expletive was "By zounds." The killer who slew men in hand-to-hand combat objected to dueling on moral grounds, and his humane treatment of prisoners of war would win universal praise.

Of French, English, and Scotch-Irish descent, the Macdonoughs of New Castle County, Delaware, were known for their staunch Anglican faith and fighting spirit. Thomas's father, a medical doctor and churchwarden, had commanded an infantry regiment during the American Revolution, and an uncle had been killed in the war. Another uncle had fought Indians on the frontier.

Most inspiring was Thomas's older brother James, who had been wounded in a sea battle during America's undeclared naval war with France. James limped back to New Castle on a peg leg, a bona fide naval hero. Fifteen-year-old Thomas eyed his brother's wooden badge of courage, listened to his tales of glory on the high seas, and decided to join the Navy.

With the help of his family's connections, Thomas entered as a midshipman, an officer trainee, at a time when the U.S. Navy was in its infancy. There was as yet no U.S. Naval Academy, no fleet maneuvers, nor even any fleet. The Navy had less than twenty full-size ships. America's "middies" received no formal training in battle tactics, so Macdonough studied the subject on his own, poring over books and journals by the masters of naval warfare, his future enemy. What he learned from the British would come back to haunt them on Lake Champlain.

Meanwhile, promoted rapidly for his heroism against the Barbary pirates, he grew into a "tall, dignified and commanding" young lieutenant:

His features were regular and pleasing. His hair and complexion were light and his eyes were blue, but the firmness and steadfastness of his look took away all appearance of the want of virile masculine energy which is often associated with a delicate complexion.

In short, he was a hunk, or so said his biographer grandson. A portrait by the American painter Gilbert Stuart isn't so flattering. It shows a pale youth with a pointy chin and nose, thin lips, dark blue eyes, and wavy red hair, a ginger kid more boyish looking than handsome. Still, aided by a natty uniform and swashbuckling reputation, he cut a dashing figure.

One who thought so was sixteen-year-old Lucy Ann Shaler, the daughter of a wealthy Connecticut businessman. Little is known about her except that she and Macdonough met in church and fell in love. But it wasn't a perfect match. At twenty-three, he seemed too old for her, and worse, too poor.

Not the type to whisk her away against her family's wishes, Macdonough bided his time and courted her with the same steely resolve he showed in battle, even taking a leave of absence from the Navy to earn money as a merchant captain. Six years after meeting, the couple finally married. There wasn't much time for a honeymoon because by that time America was at war.

Britain had been bullying her former colonies for years, seizing thousands of sailors off American merchant ships and enacting the hated Orders in Council, which restricted America's trade with Britain's enemies. Fed up, the United States declared war in June 1812.

"Our swords leap flaming from their scabbards, and cannot be returned unappeased," proclaimed a militia regiment in a letter to President James Madison. Such saber-rattling masked a hard truth: with a small, inexperienced army, a tiny navy, and state militias that

HEROES OF THE LAKES

Master Commandant Thomas Macdonough by Gilbert Stuart. (NATIONAL GALLERY OF ART.)

were mostly poorly trained and unreliable, the United States wasn't remotely ready to take on the world's greatest military power.

Before long, America found itself attacked from every direction. The British sailed up Virginia's Chesapeake Bay, marched into Washington, D.C., and burned down the White House. They also landed on the Gulf Coast and marched on New Orleans, invaded New York, and drove the Americans out of Michigan, and occupied Detroit.

The Chesapeake invasion stalled at Fort McHenry outside Baltimore, where, at dawn after an all-night bombardment, Francis Scott Key saw the star-spangled banner still waving over the ramparts. At the Battle of New Orleans, as the song of the same name goes, General Andrew Jackson's band of backwoodsmen stood behind their cotton bales, opened up with their squirrel guns, and decimated the British as they came on in high-stepping rows.

Elsewhere, America's fate would be decided on the water. South of Detroit the outcome rested with a charismatic hard charger too aggressive for his own good, and the ships he was racing to build on Lake Erie.

Oliver Hazard Perry was born with ice water in his veins. He was just two years old when he wandered into the street and was nearly trampled by a galloping horse and rider. As the hooves thundered to a stop inches away, the boy looked up and calmly asked, "You will not ride over me, will you?" The astonished rider scooped him up and brought him inside to his mother.

It was from her that Oliver inherited his fearlessness. Of Scotch-Irish descent, Sarah Alexander Perry shared a bloodline with William Wallace (hero of the film *Braveheart*). It was said that "nothing on earth could intimidate her."

By adolescence, her oldest son was captivating everyone with what an admirer called his "precocious manliness" and "personal beauty." The Episcopal bishop who confirmed Oliver was "greatly pleased by his appearance and manners." An exiled French nobleman gave the boy an exquisite watch as a token of his esteem.

Oliver excelled at everything, becoming an expert flautist, billiard player, fencer, and horseman, but his first love was the sea, a passion inherited from his father. Descended from English Quakers, Raymond Perry was a merchant ship captain and later an officer in the U.S. Navy.

The harbor town of Newport, Rhode Island, where Oliver grew up—where the Navy's towering fifty-gun frigates stopped to refit—was the perfect place to nurture his seagoing dreams. He was still just a boy when he began sailing small boats out of the harbor into Narragansett Bay. Evenings were spent under the stars learning celestial navigation. (Oliver's teacher bragged that the boy was the best navigator in Rhode Island.) His favorite boyhood pastime was leading his friends in mock sea battles fought on makeshift rafts.

After joining the Navy as a midshipman at age thirteen, the dynamic lad with a flair for the sea rose quickly: acting lieutenant at age seventeen, second lieutenant at eighteen, and, at the tender age of twenty-three, command of his own ship, the schooner *Revenge*.

Tall and burly, with a brown forelock dangling below his hairline and thick side whiskers framing his angelic features, he looked every inch the gallant sea captain, making him, to hear him tell it, irresistible to women. "You will consider it vanity when I tell you that I am in great demand here with the young ladies," he wrote to his mother while on leave in Washington, D.C. He mentioned a rich Virginia belle who liked his looks, and a certain Miss Herbert to whom he had an "attachment." He asked his mother to please discourage two women in Newport who wanted to marry him.

But his days as a ladies' man were numbered. He was twenty-one when he met Elizabeth Champlin ("Betsey") Mason, a doe-eyed fifteen-year-old with cinnamon-colored curls, the daughter of a Newport physician. Now it was the dashing young lieutenant's turn to swoon. Luckily, Betsey was just as smitten. After a four-year courtship, insisted on by her family, the couple finally wed. They would remain devoted to each other for the rest of Oliver's short life, pining for each other in love letters whenever he was away.

In 1810, a year before his marriage, his naval career skyrocketed and then came crashing down. First, off the Georgia coast, he refused to be bullied by a much larger British warship, the *Gorée*, whose captain waylaid the *Revenge* and demanded that Perry come aboard and be questioned. He was readying his crew to storm the *Gorée*, a probable suicide mission, when her captain backed down and sailed away. The incident made headlines and turned Perry into America's hero of the hour.

GLORY, GRIT AND GREATNESS

Elizabeth Champlin ("Betsey") Perry, née Mason.
(COURTESY OF THE BOSTON MUSEUM OF FINE ARTS.)

Six months later, hugging the New England coast too closely in a thick fog, he ran the *Revenge* aground on a reef. Officially exonerated (the ship's pilot had misled him), unofficially Perry took the blame. His stellar career sank with his ship.

Although he was promoted to master-commandant (in today's ranking a commander), his request for more sea duty was denied. When the war started, he was shunted off to the southernmost of America's Great Lakes, a backwater command no one wanted—Lake Erie.

Two months into the war, eager to escape the pressures and summer heat of Washington, D.C., President James Madison and First Lady Dolley Madison were heading to their Virginia estate when a rider galloped up and handed the president a message. The presidential carriage promptly swung around and made a beeline back to the White House.

Detroit had fallen. An American general had been bluffed into surrendering Fort Detroit and its garrison to a force half its size. A three-hundred-mile swath of what was then the northwest United States, from the headwaters of the Ohio River west to the Mississippi, now lay open to attack.

Panic spread, as Indians allied to the British began raiding American settlements in Ohio and Indiana. Britain wanted to turn the entire region into an Indian-occupied "buffer state" under British protection. To end the threat, Detroit had to be retaken. The first step was gaining control of Lake Erie.

That meant building a fleet from scratch in the wilds of Pennsylvania. The construction site President Madison chose, Presque Isle—the future town of Erie, Pennsylvania—on the lake's south shore, had a good harbor, plenty of timber, a foundry, a sawmill, an old cannon the locals fired every Fourth of July, and not much else. The nearest city, Pittsburgh, was over a hundred miles away. Pig iron, copper, canvas, rope, paint, tools, cannon, cannon shot, gunpowder, and dozens of other items had to be hauled to Presque Isle from great distances over narrow roads clogged with mud and tree stumps.

Leaving behind his pregnant wife and infant son in Newport to travel five hundred miles in freezing weather by mail coach, sleigh, and foot, Perry arrived in March 1813 to find himself already behind in a shipbuilding race with the British, who were almost finished with their flotilla across the lake.

Wasting no time, he charged ahead, driving his construction crews at a blistering pace, and himself even harder. His men often worked around the clock, sawing and hammering all night by torchlight, dropping where they stood to catch some sleep before starting up again. Perry went for days without sleep, dividing his time between Presque Isle and Pittsburgh, as he tried to speed up the trickle of supplies coming in from the east.

Shortages were a constant problem. Perry's head shipbuilders, Noah Brown and Daniel Dobbins, caulked the ships with molten lead rather than pitch and used wooden pegs in place of nails. Scrap metal was scrounged from the locals and melted down for bolts. Dobbins led a party nine miles across the frozen lake to scavenge materiel from an abandoned British cargo ship.

With the spring thaw, Perry's men worked in full view of the British ships circling offshore like sharks eyeing their prey. A sandbar and shore battery kept the enemy ships at a safe distance, but to prevent a raid by land, Perry resorted to an old trick. Whenever a British ship appeared, he staged a grand review, repeatedly parading the same band of militia along the shore to make it look like he had a huge force.

No attack came, and the frantic work went on all spring and into summer. Corners were cut, workmanship was hasty, and lumber was green and unseasoned, but Perry wasn't after perfection, just ships seaworthy enough for one battle.

At last, in mid-April, to cheers all around, two small gunboats slid from their blocks into Presque Isle Bay. Two more gunboats and two large brigs, *Lawrence* and *Niagara*, were launched by early July.

Earlier, in late May, braving "showers of musketry," Perry had led a land attack that broke the blockade of an American anchorage on the Niagara River, netting him a few more gunboats. He spent a backbreaking week helping to haul his new ships upriver to Lake Erie.

His fleet wasn't very big, just nine ships and fifty-four guns, but at least it was finished. Now he had to man it. The U.S. Navy sent him less than half the sailors he needed, so he combed the

Pennsylvania backcountry for volunteers—"anything in the shape of a man," as he put it. He would go into battle with farm boys, soldiers, dragoons, even invalids, crewing his ships.

Ships and crews assembled, just one problem remained, the thorniest of all: how to haul his two deep-draft brigs over the sandbar—a weeklong task—without having them blown to bits by the British ships. As Perry was mulling this over, to his astonishment the enemy flotilla disappeared. The British commander, Captain Robert Barclay, had sailed away—allegedly to see his mistress—vowing to return before the week was out.

When he did, he was the one who was astonished. Perry had outhustled him. Working nonstop, he and his men had moved the ships over the sand and into open water in just four days. Having missed a golden opportunity, Barclay withdrew.

With the end in sight, Perry was close to buckling under the strain of four grueling months. Exhausted and sick with lake fever, angry at a lack of support from his superiors, he fired off an indignant letter to Navy Secretary William Jones asking to be relieved.

Jones sent a fatherly reply meant to smooth Perry's feathers. It worked. By late summer, Perry was the same old hard charger, feeling better and firmly in command, drilling his gunners until they were ready to drop, anxious to come to grips with the enemy.

Too anxious in fact. With Perry's ships threatening the British supply line to Detroit, Barclay was forced to come out and fight. Rather than meet him on the open lake with its tricky summer breezes, Perry should have stayed inside his new base, Put-In Bay, Ohio, and let Barclay to come to him. Headstrong as ever, Perry was in no mood to wait.

At dawn on September 10, 1813, lookouts spotted Barclay's ships hull down on the horizon, their masts rising slowly with the sun toward the brightening sky. It was a scene at once beautiful and terrifying. Sixty years later, gunner John Norris from Kentucky remembered shaking so hard at the sight that his knees knocked together. His shipmate, twenty-year-old David Bunnell from Connecticut, recalled an eerie silence: "All nature seemed wrapped in awful suspense. The dart of death hung, as it were, trembling by a single hair, and no one knew on whose head it would fall."

Cries of "Enemy in sight!" sent Perry's men scampering up the rigging, and in no time his ships had left Put-In Bay and were sailing toward the enemy. As the two fleets closed, he hoisted his enormous navy-blue battle flag adorned in giant white letters with "Don't Give Up the Ship", the dying words of his flagship's namesake, Captain James Lawrence. Resplendent in the sunshine and flapping gently in the breeze, the banner roused the entire fleet to cheers. Norris ceased to shake. "My knees," he said, "were ready to do my bidding."

Just before noon, marshal music came wafting over the water. A brass band on the deck of a British ship was blasting out *Rule Britannia*. When the music stopped, the British guns opened up.

The battle would be decided by the four biggest ships armed with the lion's share of the guns, the *Lawrence* and *Niagara* for the Americans, the *Detroit* and *Queen Charlotte* for the British. On paper the fleets seemed evenly matched, except that Perry had a knife at his back he didn't know about.

Before the day was out, dying men aboard the *Lawrence* would curse the name of Master Commandant Jessie Duncan Elliot, the thirty-one-year-old captain of the *Niagara*. A war hero honored by Congress for his bravery, Perry's second-in-command was also a troublemaker who may have resented serving under a younger man. Elliot held the *Niagara* back, out of danger, leaving the *Lawrence* to crawl forward in the light breeze into the teeth of Barclay's long guns.

Before the *Lawrence* fired a shot, her hull had been splintered and her rigging shot away. Then, at close range, the British lit into her.

Canister, grape, and round shot swept her deck, dropping men like tenpins, severing arms, legs, even heads. Bunnell was spattered with the brains of the man next to him. Men blown off their feet hurtled through the air and landed in mangled heaps. Corpses sprawled on deck were torn apart as they continued to be hit. Below deck, wounded men crowded around surgeon's mate Usher Parsons faster than he could treat them. As he tended a midshipman, a cannon ball tore through the hull, slamming the boy against the bulkhead, killing him.

Emotions ran the gamut from terror, to rage at Elliot, to gallows humor. So many feathers from a burst mattress stuck to the head wounds of Lieutenant John Yarnall that he looked like a giant white owl, prompting howls of laughter.

Amid the carnage, Perry calmly strode the deck of the Lawrence, oblivious to the hail of iron whizzing around him. He issued

Above left: Captain Robert Barclay. *Above right*: Master Commandant Jesse Duncan Elliott. (BOTH IMAGES NAVAL HISTORY AND HERITAGE COMMAND, WASHINGTON, DC.)

orders with the nonchalance of a yacht skipper, even jesting at times, as if he were on a pleasure craft back on Narragansett Bay.

Once the *Lawrence* got in close, she unleashed her thirty-two pounders with deadly effect, giving nearly as good as she got, pounding both the *Detroit* and *Queen Charlotte*. Bunnell's hearing would take a year to recover.

When the ammunition ran out, Perry's gunners rammed anything made of metal down the barrels. Bunnell fired a crowbar, then grabbed a brass swivel and fired that. As his gunners fell, Perry replaced them with anyone at hand, marine sharpshooters, medical orderlies, even the chaplain. When no one was left, he called down to Dr. Parson's sick bay for volunteers. A few bleeding men staggered up to the deck and bore a hand.

By midafternoon the *Lawrence* was a shattered hulk. All her guns were knocked out and all but a handful of her crew were either dead or too badly wounded to fight. Perry had just begun to. He hauled down his "Don't Give Up the Ship" flag, rolled it up under his arm (or, legend has it, draped it over his shoulders like a cape) and, with a few oarsmen, made for the *Niagara* in a rowboat.

Gunfire churned up the water around the tiny craft, even shattering some of the oars, but what would become known as "Perry's luck" held. The boat made it through. When Lieutenant Yarnall struck the *Lawrence*'s colors, British sailors cheered, thinking they'd won the battle, until the *Niagara* was spotted heading their way bristling with fresh guns.

Perry had taken command from Elliot and steered the Niagara back into the fray. Hooked together by tangled rigging, the *Detroit* and *Queen Charlotte* were easy prey. Two broadsides later, Barclay surrendered. It was three in the afternoon. The three-hour horror was over.

HEROES OF THE LAKES

Listening in the distance, American General William Henry Harrison, the future U.S. president, anxiously awaited word of the battle. On the back of an envelope, Perry dashed off a note to him with an opening sentence that would become famous:

Dear General:

We have met the enemy and they are ours. Two ships, two brigs, one schooner and one sloop.

Yours with greatest respect and esteem,
O. H. Perry

Three weeks later, their supplies depleted, near starvation, the British regulars and Native American warriors occupying Detroit

Perry being rowed to the *Niagara,* a painting by Edward Percy Moran. (LIBRARY OF CONGRESS.)

abandoned the city and fled to Ontario, Canada, where Harrison's three thousand men routed them at the Battle of the Thames.

Let us ever remember
The tenth of September.

Poets would sing Oliver Hazard Perry's praises, and from Perrysburg, New York, to Perryville, Missouri, three-dozen counties and towns would bear his name. He had fought back from the brink of defeat to win a victory that saved the northwest United States. Crowds cheered him all along his route home. Cities honored him with parades and eighteen-gun salutes, along with a wagonload of ceremonial swords, commemorative medals, and engraved tea sets, until he grew tired of the adulation.

Home at last, "the Savior of the Northwest" settled down with Betsey and his children and quietly returned to his old duties, commanding a handful of gunboats in Newport harbor. Reflecting on the horror of Lake Erie, he told his brother, "I believe my wife's prayers . . . saved me."

Britain, meanwhile, was outraged. The pride of the world's greatest sea power had been stung. A year later, moving along another lake, the British would return with a vengeance.

In April 1814, French emperor Napoleon Bonaparte abdicated, ending two decades of war in Europe and freeing up some of Britain's best troops for action in North America.

By September, eleven thousand tough veterans of the Napoleonic Wars were marching down from Montreal, Canada into New York along Lake Champlain—the "silver dagger" pointed at

America's heartland—hoping to carve up the northeastern United States and reconquer New England.

Standing in their way were four thousand American troops, mostly poorly trained militia, and Thomas Macdonough's odd little fleet.

For the better part of two years, while the Great Lakes were getting most of the attention, the newly promoted master-commandant had been cobbling together a twelve-ship flotilla on Lake Champlain, building some ships, refitting others—including a converted merchantman, a half-built steamboat equipped with sails, and a handful of oar-powered galleys—anything that could be mounted with guns, to go with his flagship, the brig *Saratoga*.

Across the water, the British had a fleet that was about the same size until they began building the biggest ship on the lake, their flagship, the frigate HMS *Confiance*. The fleet commander, Captain George Downie, bragged that the *Confiance* alone could beat the Americans.

To have a chance against her, Macdonough needed time to build another brig, so he asked for volunteers for a daring mission. Midshipman Joel Abbot from Massachusetts would lead it. When Macdonough asked him if he was willing to die for his country, the twenty-one-year-old replied, "Certainly sir. That is what I came into the service for."

Dressed in a British uniform (a hanging offense), Abbot bluffed his way onto the British base and located the place where the *Confiance*'s spars were being stored. Later, using muffled oars in the dead of night, he and his men snuck past the sentries in a rowboat and destroyed the spars.

After three days behind enemy lines, Abbot was so exhausted he had to be hoisted aboard his ship. But his valor earned him an honorary sword from Congress and some precious time for

Commodore Joel Abbot, circa 1852.
(NAVAL HISTORY AND HERITAGE COMMAND, WASHINGTON D.C.)

Macdonough. The result was the brig *Eagle*, built by his shipwrights in an astonishing nineteen days.

In late August, as their troops prepared to march, the British challenged Macdonough to a nautical duel. They sent their respects and offered to meet his fleet on the open lake on any morning he chose. He politely declined. As he knew from his spies, the British ships were armed with long-range guns that on the open lake could pick the American ships apart. Instead, Macdonough anchored his ships end-to-end at the mouth of Cumberland Bay on the lake's western shore near the town of Plattsburgh, New York, and waited.

Eyeing the formation through his spyglass on the morning of September 11th, Captain Downie was delighted. Anchored ships were considered easy prey. Twice in the last fifteen years, at the Battle of the Nile and again at the Battle of Copenhagen, the Royal

Navy had annihilated fleets at anchor, attacking them at one end and gobbling them up piecemeal.

But Macdonough was too shrewd for that. Having studied those battles, he'd chosen Cumberland Bay because its cramped confines and fitful breezes prevented the British ships from getting around the ends of his line, forcing Downie into the kind of fight Macdonough wanted, the only kind he could hope to win, a toe-to-toe, broadside-to-broadside slugging match, in which Macdonough's powerful, short-range carronades would be most effective.

As the enemy ships rounded Cumberland Head at the bay's entrance, just one task remained. While the *Saratoga*'s crew stood at their battle stations, Macdonough knelt on the quarter deck and read aloud the "Prayer before Going into Battle" from *The Book of Common Prayer*. ("Stir up thy strength O Lord and come and help us. . . .")

Within minutes, the Americans were opening up on the *Confiance*, shooting away two of her anchors and damaging some of her spars. Ignoring the damage, Downie placed his flagship broadside to the American line, steadied her with the remaining anchors, and gave the order to fire.

Eight hundred pounds of iron from double-shotted guns slammed into *Saratoga*'s hull, rocking the seven-hundred-ton brig to her core, leaving a hundred men sprawled across her deck, killed, wounded, or stunned.

Downie had hoped to cripple Macdonough's flagship with one crushing broadside, but the *Saratoga* took the blow and shook it off. Dazed gunners picked themselves up, staggered back to their guns, and began slamming *Confiance* with broadsides of their own, and for the next two hours the fleets hammered away at each other.

Cannonballs the size of boulders turned hulls to matchwood, filling the air with splinters as deadly as buckshot. "Were you to see my jacket, waistcoat and trousers, you would be astonished how I escaped as I did," Midshipman Robert Lea wrote his mother, "for they are literally torn all to rags with shot and splinters." The mutilated bodies of less fortunate men covered the decks of both fleets, their blood draining through the scuppers like rainwater.

Aboard the two biggest ships, *Saratoga* and *Confiance*, orders were drowned out by the crash of cannon fire and the cries of the wounded, but with the ships anchored across from each other and firing as fast as possible, there was little need for orders. Both captains became gunners, bearing a hand at a cannon. Downie was crushed to death when his gun was sent flying by a direct hit. Macdonough was knocked unconscious when a boom fell on him. As he was being carried below, he woke up and went back to manning a gun, only to be knocked flat again when a severed head hit him in the face.

The cannon that crushed Captain Downie to death, on display at the U.S. Naval Academy. The dent made by the direct hit is visible at the upper left of the muzzle. (COURTESY OF PHOTOGRAPHER ALLEN C. BROWNE.)

Slowly but surely, the Americans' superior gunnery began to tell. Macdonough's gunners, drilled by him to a fine edge, poured a hundred rounds into the *Confiance*'s hull.

With victory in reach, the *Eagle*'s skipper cut his anchor cable and broke formation, leaving the *Saratoga* to bear the brunt of the enemy fire until all her starboard guns were knocked out.

At that crucial moment, the wily American played his ace. Macdonough had rigged his anchor lines in such a way that by releasing some and pulling others, he could spin his ship 180 degrees. The maneuver done, the *Saratoga* opened fire with a dozen port-side guns.

Badly crippled and taking on water, the *Confiance* struck her colors, followed soon by Downie's second biggest ship, the brig *Linnet*. Realizing the battle was lost, the smaller British ships fled. Hundreds of New Yorkers watching from shore broke into cheers.

The bloodletting over, his face and hands black with burnt powder, Macdonough became a gentleman again. When a delegation of officers from the British fleet offered their swords in surrender, he bowed and said, "Return your swords into your scabbards and wear them. You are worthy of them."

They were indeed. Nearly a hundred British officers and sailors had died that morning, with many more seriously wounded. The *Confiance*, said a medical worker, was "absolutely torn to pieces," with human remains scattered throughout the debris.

Despite their ragged gunnery, the British had put dozens of holes in the *Saratoga*. The Royal Navy might have won if Macdonough hadn't been able to spin his ship and bring fresh guns to bear.

The defeat on Lake Champlain cut the British army's waterborne supply line. The invasion force had no choice but to beat a hasty retreat back to Montreal. The northeastern United States had been saved. War would never again ravage New York's lake region.

GLORY, GRIT AND GREATNESS

News of another crushing defeat stunned the British public. To boost morale, the prime minister asked Britain's greatest soldier, the Duke of Wellington, to take command in America, but the Iron Duke begged off. "That which appears to me to be wanting in America," he remarked dryly, "is not . . . a general officer and troops, but a naval superiority on the lakes."

In peace talks going on in Belgium, the British were awaiting news of a victory that would buttress their demands for U.S. territory. When word came of the disaster in New York—what Britain's colonial secretary gingerly called the "unfortunate adventure on Lake Champlain"—the British backed off their demands, and on Christmas Eve of 1814, the two sides signed the Treaty of Ghent, which ended the war and preserved the prewar border between the U.S. and Canada. Thanks in part to Oliver Hazard Perry and

Macdonough's Victory on Lake Champlain by Eric Tufnell. The American brig *Saratoga* (*left*) engages the British frigate *Confiance* (*center*). At right is the American brig *Eagle*, which broke formation at a crucial point in the battle. (COURTESY OF THE NAVY ART COLLECTION, NAVAL HISTORY AND HERITAGE COMMAND.)

Thomas Macdonough, and the American sailors who fought and bled in two savage battles, America had survived intact.

Like Perry before him, Macdonough became a national hero. Both men wisely invested the prize money from their victories and prospered financially. Both were promoted and given command of a frigate. And both died young, Macdonough of tuberculosis at age forty-one, a few months after his beloved Lucy Ann succumbed to the same disease, and Perry on his thirty-fourth birthday of yellow fever caught during a cruise to Venezuela.

His wife Betsey never remarried. For her remaining thirty-eight years, she kept a shrine to her late husband in their home.

A heated public debate over Jesse Duncan Elliot's conduct on Lake Erie, begun while Perry was alive, would continue long after his death. The Navy convened a board of inquiry to settle the matter, but its findings, based on incomplete evidence, were inconclusive.

The War of 1812 has faded into obscurity, and with it the two young naval officers who rescued their young country. Only traces remain, scattered across the eastern United States.

Perry's "Don't Give Up the Ship" flag, all eighteen frayed square feet of it, its yellowed lettering still crisp, is on display at the Naval Academy Museum in Annapolis, Maryland.

Rebuilt almost from scratch, the *Niagara*, a giant trough with cannon and sails, is docked on Presque Isle Bay at the maritime museum in Erie, Pennsylvania.

In Plattsburgh, New York, a ten-story hike brings you to the top of the Macdonough Monument, a giant obelisk, where on a clear summer day you can look out across Cumberland Bay and, with a good pair of binoculars and a little imagination, relive the Battle of Lake Champlain.

Just as soon as you catch your breath.

Top: The Macdonough Monument, Plattsburgh, New York, looking out over Cumberland Bay. Cumberland Head is visible across the bay. (PHOTO BY F. CAVONE PRODUCTIONS/SHUTTERSTOCK.) *Bottom*: O.H. Perry's "Don't Give Up the Ship" flag. (COURTESY OF THE UNITED STATES NAVAL ACADEMY MUSEUM.)

The restored American brig Niagara. (PHOTOGRAPHER LANCE WOODWORTH AT FLICKR.COM, UNEDITED.)

CHAPTER 7

GOD BLESS AMERICA

Irving Berlin and Kate Smith

If you've never danced the Swanee Shuffle, never watched Fred and Ginger glide across the silver screen, or aren't sure how Tin Pan Alley got its name, you may not be familiar with America's greatest songwriter.

During a career that stretched from the time of hand-cranked Victrolas to the era of hard rock, Irving Berlin, the piano-plunking refugee from Russia, wrote catchy dance numbers and mournful ballads, waltzes, rags, lullabies, serenades, novelty tunes, and musical scores—thousands of songs in all, with dozens of hits, including the all-time best-selling single, the Oscar-winner "White Christmas."

But he is best known for a patriotic song that can still move Americans to rise to their feet and sing. The title was a saying his mother liked to quote. The opening notes came from an old vaudeville tune.

Berlin thought the song was too corny to be published. "God Bless America" would lie buried among his rejects for twenty years, unsung and all but forgotten, until just the right moment: when war clouds started gathering over Europe.

And the perfect voice came along.

Kate Smith was blessed with a voice like a Cartier diamond, radiant and flawless, five pitch-perfect octaves tender enough to bring people to tears, powerful enough to make them stand up and cheer.

She never took singing lessons. She didn't need them. Singing came as naturally as breathing. Still in diapers, before she could talk, she would toddle over to where her mom was playing the family piano and try to sing along.

By age nine, during World War I, she was singing at war bond rallies, a plump little girl inspiring people to open their wallets for Uncle Sam, something she would do to even greater effect two decades later.

By her mid-teens, she was blowing away the competition at talent contests and singing at the White House for President Warren Harding. She basked in the adulation and dreamed of being a star. Too burly to be a glamour queen, she was determined to sing her way to the top.

Kathryn Elizabeth Smith, the future "First Lady of Radio," who would turn an Irving Berlin reject into America's most beloved song, always claimed she was born in Greenville, Virginia. Greenville had a nicer ring to it than her real birthplace, Washington, D.C., where she was born May 1, 1907, to a staunch Presbyterian mother and an equally staunch Catholic father, a newspaper distributor, neither of whom wanted a show-business career for their daughter. They wanted her to be a nurse and sing alongside her father in the church choir. When Kate refused, her father slapped her. "We love you Kathryn," he told her. "We want to protect you from misery and degradation."

GOD BLESS AMERICA

She tried nursing school, then secretly quit when she landed a week-long spot in a vaudeville show. She was "praying for a miracle," she would recall, hoping to be discovered before the week was out.

On the last night it happened. A Broadway producer offered her a part in his new musical, *Honeymoon Lane*. It was just a bit part, but when she sang the heart-tugging ballad "The Little White House" (*"It's a little white house with little green blinds at the end of Honeymoon Lane"*), the audience went wild. The *New York Times* hailed Kate as the "newcomer who stopped the show."

Soon, with her parents blessing, she was starring on Broadway and turning out hit records. Still in her early twenties, Kate had fulfilled her dream of being a star.

Yet, walking along the Potomac River during a visit home, she considered drowning herself. She couldn't bear being mocked on stage about her weight. When a costar would say of her, "When she sits down, it's like a dirigible coming in for a landing," the audience that had cheered Kate's singing would burst out laughing. The insults left her sobbing in her dressing room.

She was considering quitting show business when her manager convinced her to try a new medium that was on the rise in the days before television. On radio, her physique wouldn't matter. Only her incomparable voice.

The night her fifteen-minute radio show debuted, when she sang "When the Moon Comes Over the Mountain," her first trademark song, the CBS switchboard lit up, an operator said, "like the sky on an old-fashioned Fourth of July."

Thus began a twenty-year run on the radio. Celebrities lined up to appear on her Thursday-night variety show, *The Kate Smith Hour*. Baseball great Babe Ruth did comedy sketches. Hollywood stars did readings from their latest movies. Abbott and Costello introduced their legendary routine "Who's on First?"

But the main attraction was Kate herself. People loved her singing. Her voice wasn't sultry like Peggy Lee's or bluesy like Billie Holiday's, but she was a born crooner—"The Song Bird of the South"—the female version of Bing Crosby or Perry Como, with more range and power. No one could belt out a song better than Kate Smith.

She cut a score of gold records—ranging from sentimental standards like "The White Cliffs of Dover" to the chirpy jingle "The Woodpecker Song"—and drew record crowds at New York's prestigious Palace Theater. The night she appeared at Madison Square Garden, five thousand people had to be turned away.

Meanwhile, by day, she became the queen of talk radio. Airing weekdays at noon, *Kate Smith Speaks*—fifteen minutes of headline news, feel-good stories, flag waving, and homespun advice—became the country's leading daytime radio show, turning the popular singer into a cultural icon.

Kate Smith, circa 1928. (GETTY IMAGES.)

GOD BLESS AMERICA

A typical show might include a report on the famous Dionne quintuplets ("The five little Donnies are all abed today with sore throats"), the story of "The Star-Spangled Banner," a plug for the Memphis Cotton Carnival, and a breezy lecture on gardening ("My African violets are blooming their heads off"), before she signed off with her homey tagline: "Thanks for listenin' and goodbye folks."

The material wasn't all fluff. She spoke out against age discrimination, trumpeted the achievements of American women, and condemned the terror of the Kristallnacht in Nazi Germany. The "home-grown Reds" who paraded in New York City on May Day were "hammer-and-sickle fanatics."

The people she called "plain Mr. and Mrs. America" loved the show and adored its star. Ten million listeners a day tuned in. Fan mail poured in, three thousand pieces a day. That a staff of writers scripted every word she spoke didn't lessen her star power. Feted wherever she went, she gathered tributes like some people collect coins: "Queen of the Air" in New York City; honorary member of the Texas Rangers; honorary citizen of Tennessee; honorary member of the Sioux Nation.

At the height of her fame, polling showed she was the third most admired woman in the world, behind Eleanor Roosevelt, ahead of the queen of England.

Most amazingly, in a show-business world of philanderers, drunks, and phonies, Kate was as wholesome as she sounded on the air, a demanding but pious woman who refused to sing racy lyrics, lavished time and money on worthy causes—wounded vets, sick children, needy families—and abstained from alcohol and sex.

She never married and, for all anyone knows, never had a serious romantic relationship. Whispers of an affair with her longtime manager, Ted Collins, a married womanizer, were unconfirmed and probably untrue. The reason she gave for staying single was that she was married to her career.

Kate Smith: In 1943 (*top*), and 1965 (*bottom*). (BOTTOM PHOTO GETTY IMAGES.)

Her sole weakness was comfort food. To ease the stress of performing, she feasted on seven-course meals and binged on sweets. Breakfast was often a box of donuts. A cake—not just a slice, a whole cake—awaited her after every show. She downed as many as six milkshakes at a sitting, and for a midnight snack ate ice cream from the carton.

GOD BLESS AMERICA

After years of roller-coaster dieting, she learned to take her two-hundred-fifty-pound girth in stride. She published her *Company's Coming Cookbook* and titled her first autobiography *Living in a Great Big Way*. Asked about her weight, she would smile and wink and give a figure in the low two hundreds.

Five feet ten inches tall, long-legged and broad shouldered, she wore her weight well. Draped in her signature black gown, head held high, shoulders thrown back, her arms outstretched in a gesture of embrace as she performed a patriotic song, she seemed as grand as Lady Liberty herself. The only thing missing was the torch.

Even the president saw Kate that way. At the White House, when Franklin Roosevelt introduced her to the king and queen of England he reportedly said, "Your Majesties, this is Kate Smith. This is America." Lady Liberty aside, no one typified America more than Kate Smith.

Except maybe a dark-eyed little genius from Belarus.

Born Israel ("Izzy") Baline on May 11, 1888, in what was then czarist Russia, the man the world would know as Irving Berlin was four years old when he watched his house being put to the torch.

This earliest and ugliest memory happened during an anti-Semitic reign of terror, and drove his family to join an exodus of biblical proportions, the flood of Russian Jews fleeing halfway around the world to America.

Speaking no English, taking only what they could carry, the Balines traveled eleven hundred miles, by train, horse cart, and foot, to Belgium, crossed the Atlantic Ocean in a squalid passenger ship, and ended up, almost penniless, in New York City's worst slum, what one writer called "a gray stone world of tall tenements," known more crudely as Jew Town: Manhattan's Lower East Side.

GLORY, GRIT AND GREATNESS

It was said to be the world's most crowded neighborhood, little more than a square mile housing a quarter million people, mostly poor immigrants, jammed together on rat-infested streets that stank of garbage and raw sewage.

The family of eight Balines, along with a lodger they took in, squeezed into their new home, a three-room tenement apartment. Izzy, the baby of the family, often slept on the fire escape. The family laundry hung from a clothesline that stretched from building to building high over the street. The bathroom was an outhouse out back.

Yet, to the boy who'd seen his home in Belarus being burned down by terrorists, life in a New York slum was a blessing. "Everyone should have a Lower East Side in their lives," he said looking back. "I wasn't cold or hungry. There was always bread and butter and hot tea. I slept better in tenement houses than I sleep now in a nice bed."

Despite no musical training and a thin, raspy voice, he dreamed of being a singer. Back in the old country, his father, grandfather,

Manhattan's Lower East Side, circa 1909. (LIBRARY OF CONGRESS.)

GOD BLESS AMERICA

and great-grandfather had all been cantors in the synagogue. Crooning was in his blood.

Izzy was fourteen when he quit school and left home to chase his dream. He became a busker, singing for tips on the streets and in the saloons and brothels of two of Lower Manhattan's seediest neighborhoods, China Town and the Bowery, serenading half-drunk sailors and hookers, amusing them with racy lyrics he improvised. On a good day he earned enough for a bed in a flophouse (fifteen cents for the bed, sheets were ten cents extra), and a steak dinner (twenty cents) in one of the eateries along the Bowery's Park Row, with a nickel or two left over for the rocking chair he wanted to buy his widowed mother.

He was eighteen and working as a singing waiter when he sat down at the piano one night after the bar closed to try to plunk out some tunes that were rattling around in his head. Thus began a passion for songwriting that would last eighty years.

The twentieth century's first decade was a good time to be a songwriter, a time of lavish Broadway musicals and barrel organs on city streets, when scandalous dances like the bunny hug and turkey trot (banned in Boston) were all the rage, and every parlor seemed to have a piano, or at least a new-fangled windup record player called a Victrola.

Classical composers, hack musicians, anyone with music to peddle, flocked to where New York's music companies had set up shop, a row of shabby walkups shoehorned into a single block of Midtown Manhattan. Someone who thought all the clanking pianos sounded like pots being banged together christened the street. For fifty years Tin Pan Alley would produce the lion's share of the country's hit songs, including some of the biggest by the man who now called himself Irving Berlin.

Easier on the ears than Israel Baline, his pen name first appeared on the sheet music of his first published song, a forgettable ditty called "Marie from Sunny Italy." The royalties were thirty-seven cents.

Irving Berlin age 18.

He kept composing, humming his melodies to a professional musician who put them down on paper. Eventually, Berlin taught himself to compose on a special piano equipped with a lever that allowed him to switch keys, but he would always need a transcriber. Berlin once asked the famous operettist Victor Herbert whether he should take music lessons. Perhaps not, was the reply. They might cramp Berlin's natural style.

It was a style born of monomania. A high-strung insomniac who thrived on chewing gum and cigarettes, he routinely worked almost around the clock, writing songs all night, pacing and fidgeting, tinkering with them for hours and then scrapping them and starting over, an ordeal he likened to sweating blood: "When the drops that fall off my forehead hit the paper, they're notes."

He worked on songs in taxis and restaurants and while shaving—jotting down lyrics and humming out melodies—and brought his key-switching piano with him on vacation so he could compose in hotel rooms and on cruise ships. He perfected his craft by writing thousands of songs that either weren't published or were outright flops. Once, when someone congratulated him on all his hits, he blurted, "There's no one who has written so many failures."

Dually gifted, he wrote his own words *and* music. He said his aim was to touch America's soul with simple lyrics—"easy to sing, easy to say, easy to remember"—and melodies meant to be whistled.

The result was enough hit songs to fill a jukebox, nearly 150 for his career, about everything from a singing chimp ("That Monkey Tune"), to jet setters in hell ("Pack Up Your Sins and Go to the Devil"), to dancing waiters with flat feet ("Swanee Shuffle"), not to mention classics like the Oscar-nominated "Cheek to Cheek," the song Fred Astaire and Ginger Rogers danced to in their hit film *Top Hat*.

Berlin's breakout year was 1911, when the promising twenty-three-year-old with some now-forgotten hits to his credit wrote his first blockbuster, the song *Variety* called "the musical sensation of the decade." "Alexander's Ragtime Band" would become a global phenomenon translated into dozens of languages, and make Irving Berlin world famous.

Overnight the former singing waiter became "The King of Ragtime." He bought his mom a house to go with her rocking chair, moved into a plush uptown apartment, and joined New York's high society, the glamorous world of "High hats and Arrow collars/White spats and lots of dollars," he would immortalize, and poke fun at, in another standard, "Puttin' on the Ritz."

The Friars Club, New York's toniest show-business fraternity, made him a member, and he began composing for vaudeville's hottest revue, *The Ziegfeld Follies.* When he appeared in a Times Square theater to sing a medley of his songs, women swooned.

He ignored them because he was already smitten. Dorothy Goetz was an aspiring singer, a shapely twenty-year-old brunette as driven as her future husband. He'd first laid eyes on her in his office when she got in a hair-pulling fight with another woman who wanted to debut one of his songs. The other woman got the song; Dorothy got the writer. Theirs was a storybook courtship, followed by a marriage cut short by tragedy. On their honeymoon in Havana, she caught typhoid fever and died soon after.

He poured out his grief in the million-seller "When I Lost You," and buried himself in his work, writing scores for Broadway shows, becoming part owner of a New York theater, the Music Box, starting up a music-publishing house, and getting fabulously rich.

It wasn't until fourteen years after Dorothy's death that he published his famous ode to rediscovered love, "Blue Skies":

Blue skies smiling at me
Nothing but blue skies do I see. . . .
Blue days, all of them gone
Nothing but blue skies from now on.

Sinatra. Judy Garland. Ella Fitzgerald. Even British rock star Rod Stewart. Everyone who was anyone would cover the song. Willie Nelson topped the country charts with it fifty years after it was written.

Berlin wrote "Blue Skies" in 1926 after marrying again. His new bride was a twenty-two-year-old journalist from Long Island, fifteen years his junior. Blond, blue-eyed Ellin Mackay, said *Variety*, was "one of the most beautiful women known to the society pages."

Their courtship had kept tongues wagging, and tabloids speculating, for months. The workaholic songwriter was in no hurry to propose, which was fine with Ellin's anti-Semitic father, the domineering telegraph and mining tycoon Clarence Hungerford Mackay. "My daughter will marry Irving Berlin only over my dead body," he vowed to the press. When Clarence dragged his daughter off to Europe, Berlin wooed her back with a song that would become another classic, "Always."

With Clarence threatening to disown his daughter, the couple eloped and were married at city hall amid a swarm of reporters. "We were married and we are very happy and that's all we care to say," Berlin told them. He might have added that fourteen years of loneliness were finally over.

The marriage would last sixty-two mostly happy years. The father-daughter feud would thaw after Berlin bailed his father-in-law out of financial trouble during the Great Depression. Ellin would

Newlyweds Irving Berlin and the former
Ellin Mackay were hounded by the press.

become a best-selling novelist, and she and Irving would endure the loss of an infant son and raise three daughters together. She would patiently bear her husband's short temper and bouts of depression, and the couple would remain devoted and faithful to each other until Ellin died in 1988 at age eighty-five, a year before her husband.

Irving Berlin once said of America, "Everything I have and everything I am I owe to this country." In 1918 he expressed his feelings in a song he wrote for a musical revue. The song's title was a phrase his mother taught him. The melody may have come from a vaudeville tune called "When Mose with His Nose Leads the Band." Berlin's song didn't work as a show tune, so he scrapped it.

Twenty years later, with Nazi troops on the march and war looming over Europe, he dug up the song and reworked it. He considered the new version too sentimental for publication. "God Bless America" might have ended up back on the scrap heap if Kate Smith hadn't asked for something patriotic to sing for Armistice Day.

The night she debuted the song on *The Kate Smith Hour*, November 10, 1938, as millions tuned in, she sang the introduction, seldom heard today, softly and slowly, like a love song—

While the storm clouds gather far across the sea,
Let us swear allegiance to a land that's free
Let us all be grateful that we're far from there
As we raise our voices in a solemn prayer.

—before launching into the now-famous chorus:

God bless America
Land that I love

GOD BLESS AMERICA

Upping the tempo to a march, backed by a full orchestra, she belted out the lyrics in her stirring contralto voice, equal parts joy, reverence, and raw power, building to a ringing finale, a crescendo of the words "home sweet home" that threatened to shake the rafters.

Afterward, as the studio audience cheered, she waived Berlin up to the stage and gave him a bear hug that lifted the diminutive songwriter off his feet.

They knew they'd done something special. "I wonder if I'll ever receive as great a thrill as I did last Thursday night," wrote a columnist for the *New York Sunday Mirror*. "Telephones began to ring with messages from all parts of the country asking, 'Where can we get that song Kate Smith just sang?'"

Critics called the song corny, nauseating, "chauvinistic babble" a "trashy ballad." Clergymen denounced it for trivializing religion. The Ku Klux Klan denounced it because Berlin was Jewish. Songwriter Woody Guthrie got so sick of hearing it he composed his famous ballad "This Land Is Your Land" in protest.

Yet, for many Americans "God Bless America" became a cherished hymn to their beloved country, a source of strength in trying times, as reassuring as Mount Rushmore. Or Old Glory snapping in the breeze.

People couldn't get enough of the song. They not only wanted to hear it—Smith's recording of it would spend fifteen weeks on the Billboard charts—they wanted to sing it. Written in Berlin's simple style (*easy to sing–easy to say–easy to remember*), with a catchy melody and memorable lyrics, the song was tailor-made for sing-alongs. "If you can't sing 'God Bless America,'" a journalist quipped, "you can't sing anything."

Groups from the Daughters of the American Revolution to the American Labor Party opened their meetings with it. Crowds big and small sang it in churches, synagogues, and schools; at ballgames during halftime and the seventh-inning stretch; and at both the 1940

Kate Smith and Irving Berlin with servicemen in 1943. (GETTY IMAGES.)

Democratic and Republican conventions, often while standing at attention with their hats off, an honor usually reserved for "The Star-Spangled Banner."

An effort arose in Congress to make "God Bless America" the national anthem, which failed after Berlin and Smith objected out of respect for "The Star-Spangled Banner." Berlin received a Congressional Gold Medal for "his services" in writing "God Bless America," services that included donating all the royalties in perpetuity to the Boy Scouts and Girl Scouts of America.

It is now May 16, 1976. Thirty-eight years have passed since that November debut. In America's bicentennial year, "God Bless America" remains the country's unofficial national anthem. Tonight at the Spectrum Arena in Philadelphia, Kate Smith will sing it live for the last time.

GOD BLESS AMERICA

She has performed the song hundreds of times over the years, transforming the popular singer and radio personality into a symbol of national pride. During the war years in the 1940s, her performances moved audiences to tears, helping her sell a record $600 million in war bonds.

In the rock-and-roll era that followed, her fans stayed loyal, swamping NBC with four hundred thousand protest letters when her TV variety show was cancelled. Well into the 1960s she was still cutting million-selling albums and packing Carnegie Hall.

Since 1969, her rendition of "God Bless America" has been a good-luck charm and fan favorite at the home games of the Philadelphia Flyers hockey team. Her swan song will take place at the Stanley Cup Finals between the Flyers and the Montreal Canadiens.

The arena goes dark, an organ fanfare strikes up, and she enters the rink on a red carpet under a spotlight, waving and blowing kisses. The crowd erupts. They love her here. As she's done since she was a child, she smiles and bows, squares up to the mic, throws her head back, and starts to sing.

Sadly, fifty years after she wowed 'em in *Honeymoon Lane*, time has finally caught up with the once-robust sixty-nine-year-old. A lifelong sweet tooth has left her with diabetes and heart trouble. Arthritis has gnarled her hands. Thick glasses and bloated features—the result of radiation treatments for cervical cancer—make her almost unrecognizable. She sometimes has trouble remembering lyrics. She will live ten more years but will spend the lion's share of them wheelchair-bound, and when President Ronald Reagan honors her with the Medal of Freedom a few years from now, she will sit silently and uncomprehendingly through the ceremony, in the grip of dementia.

But tonight, despite it all, magic happens. The flesh is weak, but the voice is still vibrant. Like an old, out of shape Babe Ruth blasting a ball clean out of Forbes Field in one of his last at bats, she nails her most famous song one last time—crushes it.

By the end, fifteen thousand cheering hockey fans are on their feet. She doesn't linger to revel in the applause. For an old trouper it's all in a day's work. Besides, skaters are starting to warm up. There is a championship game to be played.

She takes a quick bow, clasps her hands over her shoulders in a victory salute, and calmly walks out of the spotlight to a thunderous ovation, a final triumphant exit for the plump little girl who dreamed of being a star.

From the time she first sang in public at age nine, Kate Smith's career lasted an astonishing sixty years. Irving Berlin's would last sixty-four.

Kate Smith waves to the crowd in Philadelphia's Spectrum Arena before singing "God Bless America" for the last time in public. (GETTY IMAGES.)

GOD BLESS AMERICA

He began singing for pennies at age fourteen, the year before the Wright brothers' first flight, and wrote his final showstopper, "An Old Fashioned Wedding," for a revival of *Annie Get Your Gun*, three years before the moon landing.

From the early 1930s, through the Great Depression and World War II and into the 1950s, he wrote the scores for more than a dozen musicals for stage and screen, and received seven Oscar nominations for best song.

Among the nominees was a "cockamamie little" Christmas carol from the 1942 film *Holiday Inn*, a throwaway song no one paid much attention to until homesick GI's began flooding Armed Forces Radio with requests to play it. "White Christmas" would not only win an Oscar, Bing Crosby's version would become the best-selling single ever, with sales of fifty million and counting.

Berlin's wartime Broadway smash *This Is the Army* featured a three-hundred-soldier cast, a forty-four-piece orchestra, lively song-and-dance numbers, and the composer's own high-pitched croak. His raspy rendition of his song "Oh How I Hate to Get Up in the Morning," a musical bellyache about a barracks bugler ("I'll amputate his reveille and step upon it heavily."), left audiences in stitches. He donated the show's seven-figure proceeds to the Army Emergency Relief Fund.

Traveling with the cast and crew in a ramshackle Dutch freighter, braving enemy submarines and air raids, along with sweltering heat, bad food, torrential rains, and the crack of rifle fire echoing from the jungle, he staged *This Is the Army* for frontline troops from Naples to the Philippines, earning him a Medal of Merit from President Harry Truman.

In an era rife with segregation, Berlin saw to it that the cast included two dozen African Americans, making the show's theatrical troupe America's only integrated military unit.

Irving Berlin entertaining the crew of the USS Arkansas, 1944. (ALAMY.)

By the war's end, the songwriter had become a marquee name. *Irving Berlin's Easter Parade*, one of the highest grossing films of 1948, was topped six years later by *Irving Berlin's White Christmas*, which set box office records for a Hollywood musical.

But the heights he rose to made his inevitable fall all the worse. By the 1960s, Elvis and the Beatles were topping the charts, cheek-to-cheek dancing and big-band music were ballroom relics, and "the greatest songwriter of all time" (Cole Porter's words) had written his last hit. After its advance ticket sales ran out, Berlin's last Broadway musical, *Mr. President*, sank like a stone, and MGM scrapped his final film project, *Say It with Music*.

Disillusioned, he spent the last twenty of his one hundred and one years as a virtual recluse (a "reclusive immortal" the *New York Times* called him), holed up with Ellin in their Catskills estate and Manhattan townhouse, writing songs that were never published, and serenaded every Christmas Eve by a group of fans singing "White Christmas" under his window.

GOD BLESS AMERICA

One bitterly cold night he invited them in to warm themselves. The winner of three presidential medals and some of music's most coveted awards told them, "That's the nicest Christmas present I've ever had."

When he died in 1989, his obituaries quoted high praise from another famous composer, Jerome Kern: "Irving Berlin has no place in American music. He is American music."

Twelve years later, after the 9/11 terrorist attack, the song Berlin considered too sentimental helped rally the nation. "After the Terror: God Bless America" proclaimed the cover of *Newsweek*. Berlin's anthem seemed to crop up everywhere, proudly sung by teary-eyed traders at the stock market's opening bell, by the glitterati at a New York gala ("For [once] it didn't really matter what people were wearing," gushed an amazed Oscar de la Renta.) and by a bipartisan chorus of Congressmen and women on the Capitol steps. The commissioner of baseball ordered it sung at every major league game.

The patriotism triggered by 9/11 would fade, but Berlin's song would be kept alive by some gifted artists who put their stamp on it. Whitney Houston did a blues rendition at Radio City Music Hall, and Celine Dion's torchy version was a top-twenty hit. Most famously, tenor Ronan Tynan crooned the song in his Irish brogue at New York Yankee and Boston Red Sox games. Yet, when it comes to picking the definitive version of "God Bless America," it's no contest. To this day no one can touch Kate Smith.

"It's one of the most beautiful compositions ever written," she told her listeners on that long ago November night. "A song that will never die."

She may be right, as long as there is an American flag flying overhead and a crowd of people below willing to sing their hearts out.

CHAPTER 8

THE YEGG HUNTERS

The Lawmen Who Took Down John Dillinger, Baby Face Nelson, Pretty Boy Floyd, and Machine Gun Kelly

On the afternoon of November 27, 1934, near the town of Barrington, Illinois, forty miles north of Chicago, a manhunt was underway.

The FBI was combing the area for America's "Public Enemy Number One," Lester Joseph Gillis, better known as Baby Face Nelson. Wanted for a string of bank robberies and murders, he was reportedly headed toward Chicago in a stolen Ford with his sidekick and fellow gangster, Johnny Chase.

On a lonely stretch of highway, FBI agents Sam Cowley and Herman Hollis spotted the stolen Ford coming the other way. The agents made a tire-screeching U-turn and sped up in pursuit.

In keeping with the wholesome image the FBI prided itself on, Cowley and Hollis were young, clean-cut family men. Cowley, thirty-four, was a devout Mormon from Idaho with a wife and two young sons. Hollis, a thirty-one-year-old Catholic from Iowa, had a wife and four-year-old son. Both agents had law degrees and dreamed of someday opening a law practice.

Neither had any illusions about the man they were chasing. Baby Face Nelson was a trigger-happy psychopath who had already murdered an FBI agent, executing him while he sat in his car. Wherever Baby Face travelled, he brought along a small arsenal of firearms. It was doubtful he could be taken alive.

Normally, he would have floored the gas pedal and tried to outrun the agents, but the Ford's engine, damaged during an earlier shootout, had begun smoking and sputtering. The engine died just as Nelson was turning down a gravel road.

Hollis slammed on his brakes and the FBI car skidded to a halt just off the highway, thirty feet from the crazed killer.

The hunt for Baby Face Nelson was but one episode in a federal anticrime crusade begun a year and a half earlier on a sunny June morning in 1933 outside a train depot in Kansas City, Missouri.

Half a dozen federal agents and local lawmen were escorting a fugitive, a bank robber named Frank Nash, across the parking lot on his way back to prison when someone yelled, "Let 'em have it!" Three or four unidentified gunmen—witnesses weren't sure how many—sprang up from behind some parked cars, opened fire with automatic weapons, and sped off in a getaway car.

"Officers Are Mowed Down by Gangsters" blared a *New York Times* headline. What soon became known as the Kansas City Massacre left three police officers, one federal agent, and Nash himself dead, and the country in shock. Gang violence was nothing new in America—mobsters in New York City and Chicago had been gunning each other down for years—but the slaughter of four law officers demanded action. Within days of the massacre, President Franklin Roosevelt's administration publicly declared a "War on Crime."

Declaring the war was easy. Winning it wouldn't be.

On one side were the gangs of Tommy-gun-toting outlaws—"yeggs" they called themselves—who were rampaging across the Midwest, robbing banks, kidnapping wealthy victims, and gunning down cops.

Led by men with flashy nicknames—Baby Face Nelson, John "Gentlemanly Johnny" Dillinger, Machine Gun Kelly, and Pretty Boy Floyd—these outlaws were as clever as they were ruthless. They used hostages as human shields, drove armor-plated getaway cars, and mapped out escape routes covering hundreds of miles, with cans of gasoline stored along the way. While on the lam, they had a network of allies—family, friends, underworld figures—who helped them hide out.

On the other side was what one critic called "an odd-job detective agency." When the War on Crime began, the Federal Bureau of Investigation was a fledgling group of mere "fact finders" who handled lesser crimes, like interstate auto theft. Scoffed at by big-city police, FBI agents weren't authorized to make arrests or carry firearms. On raids, armed only with bats, agents had to wait outside while pistol-wielding cops went in first.

As the war heated up, the FBI was transformed into a full-fledged national police force, with dozens of new recruits, arrest powers, and plenty of firepower. But the sweeping changes outpaced the bureau's training program, which would lead to mistakes.

Lacking training, FBI agents would rely on raw courage and, above all, sheer doggedness. No matter where the yeggs hid, the FBI would never be far behind. Agents like Melvin Purvis, the head of the Chicago office, would prove as relentless as foxhounds.

GLORY, GRIT AND GREATNESS

Up against murderous gangsters, he and his men were quick on the trigger. Too quick on a frigid April night in 1934 in the Wisconsin woods.

The tragic chain of events began with a hot tip. Purvis got word that the celebrity bank robber John Dillinger and his gang, including the notorious killer Baby Face Nelson, were staying at the Little Bohemia Lodge in northern Wisconsin. Purvis got bogus tips all the time. He had a hunch this one was real.

Two days earlier, six sinister-looking men had arrived at the lodge with some of their wives and girlfriends for a weekend getaway. The men's luggage weighed a ton—"What are these guys, hardware salesmen?" blurted the bellhop—and they had bulges in their coats, sure signs they were carrying firearms. Suspicion turned to alarm

FBI Agent Melvin Purvis. (ASSOCIATED PRESS.)

when the lodge owner saw a picture of one of the men on the front page of the morning paper. He was John Dillinger. The mystery guests were the Dillinger gang. Fearing for the safety of her eight-year-old son, the owner's wife alerted the FBI via a hastily scribbled note smuggled out of the lodge in a pack of cigarettes.

Near midnight on a moonless Sunday, Purvis and sixteen other FBI agents began creeping up on the lodge just in time to see three shadowy figures with rifles emerge onto the front porch. The three men entered a car and sped headlong down the long gravel driveway, ignoring repeated shouts of "Stop the car! Federal agents!"

Purvis and another agent yelled "Fire!" and by the time the bullet-riddled car lurched to a halt, all three occupants had been shot; one of them, a government worker named Eugene Boisneau, was dead. The men were innocent locals who had apparently been out hunting and had just dined at the lodge. Driving with a blaring radio, possibly drunk, they didn't hear the orders to stop.

As the fatal error dawned on Purvis, gunfire kicked up the gravel at his feet. A member of the Dillinger gang was shooting at him from the shadows. Dillinger himself joined in with a Tommy gun from a balcony. Before the agents could regroup, he and his gang escaped.

But the bloodshed wasn't over. Later that night, a mile away, a small young man walked up to the open driver's window of a car containing two FBI agents and a local constable. "I know you bastards are wearing bulletproof vests so I'll give it to you high and low," the man snarled before blasting away with a pistol at point-blank range. As the shooter fled, the constable and one agent lay bleeding. The other, the affable, soft-spoken Agent Carter Baum, the father of two baby daughters, lay dead. Baby Face Nelson had killed his first cop.

The press too moved in for the kill. The *Chicago Times* called the hunt for Dillinger a "farce-comedy ... turned to tragedy." FBI agents were "comic opera cops," declared the *Milwaukee Sentinel*. "Dillinger is going to get in accidentally with some innocent bystanders some time," quipped humorist Will Rogers, "then he will get shot."

Purvis tendered his resignation, but FBI director J. Edgar Hoover, the young station chief's friend and mentor, refused to accept it. Even so, Eugene Boisneau and Carter Baum were dead, the bureau was being raked over the coals, and John Dillinger and Baby Face Nelson were free to continue their crime spree.

A year into the War on Crime, the yeggs were winning.

In between the Kansas City Massacre and the shootout at Little Bohemia, there was another headline-making shocker. This one in Oklahoma City.

On a Saturday in July, 1933, wealthy oilman Charles Urschel was enjoying a late-night bridge game in his sunroom when two armed gunmen calmly walked in, one saying, "Keep quiet or we'll blow your heads off."

While Urschel's companions watched in horror, the men hustled him to a car, put tape over his eyes, shoved him in the back seat, and sped away. Soon, anonymous instructions arrived by mail. Urschel's wife paid the largest ransom ever up to that time, $200,000, and after nine days in captivity, weak and hungry but unharmed, Urschel was released.

J. Edgar Hoover publicly called the kidnappers "sewer rats." To hunt them down, he turned not to one of the young college graduates who were common in the FBI, but instead to a fifty-year-old high school dropout, a squat, bespectacled former Texas ranger who carried a six gun on his hip and wore a ten-gallon hat.

The career of Gustave Tiner Jones from San Angelo, Texas, was the stuff of frontier lore. Before becoming the FBI's San Antonio station chief, Gus Jones chased gun runners on the Mexican border, went undercover to thwart a gang of train robbers in the New Mexico territory, and kept the peace on the streets of a wild Texas town. He was renowned in the bureau for tracking down a car thief who murdered an FBI agent.

Now, Hoover handed Jones another tough nut to crack. The Urschel kidnapping was a mystery without any clues. Kept blindfolded throughout his captivity, Urschel had no idea who his abductors were or where they held him. Urschel himself thought the case was hopeless, like looking for a needle in a haystack, he told Jones.

Jones sat the oilman down and began tapping into his amazing memory, gleaning everything he'd heard over nine days. At a gas station stop in the getaway car, someone had mentioned "broom corn," suggesting southern Oklahoma, where that crop was grown. Later, the car sounded like it was rolling over a wooden bridge, most likely the bridge that crossed the Red River into Texas. At Urschel's place of captivity, pigs squealed and a rooster crowed, suggesting a farm.

Left: FBI Agent Gus Jones. *Right*: Jones in his Texas Ranger days.

Urschel had also heard daily flights overhead, and a rainstorm the day before his release. "Now I want you to think hard, Mr. Urschel," Jones said. "This may be the most important question of all." Had Urschel heard the flights the day of the storm? He closed his eyes and thought for a moment. No, he hadn't. That became the clincher. Airline logs, including an irregular one on the day of the storm, pointed to Wise County, Texas. "The haystack," Jones said, "has just gotten a lot smaller."

Meanwhile, a lucky tip came in from the Fort Worth, Texas, police. A detective had seen a Cadillac parked in Fort Worth with red Oklahoma dirt on the wheels and an Oklahoma newspaper reporting the kidnapping on the front seat. Urschel said the kidnappers drove a big car, possibly a Cadillac. The one the detective saw belonged to a woman named Kathryn Kelly whose stepfather owned a farm near the town of Paradise, Texas—in Wise County. An undercover FBI agent visited the farm. It matched Urschel's recollections to a tee.

In the raid that followed, the FBI nabbed the two men who had guarded Urschel: the farm's owner, R. G. Shannon, and his son. Gus Jones saw a bank robber he knew named Harvey Bailey napping on a cot outside. Jones brushed Bailey's nose with the barrel of a Tommy gun and woke him up. Bailey eyed the .45 lying by his hand, looked down the barrel of the Tommy, and gave up with a shrug: "Hell, a fella's gotta sleep some time." In his pocket were $700 in marked bills from the ransom.

Agent Jones had cracked the case, but it wasn't over. Still at large was the key figure in the kidnapping, the yegg with the coolest nickname of all.

The source of the nickname was his wife, who liked to brag that her husband could shoot walnuts off a fence with his submachine gun.

THE YEGG HUNTERS

A former small-time bootlegger, George "Machine Gun" Kelly took up bank robbery so he and his wife Kathryn ("Kit") could live in style. Shortly after they met, armed with a Tommy gun she bought him, George joined a gang of thieves he'd met in prison and began robbing banks from Texas to Minnesota.

His share of the take ran to five figures per robbery—a small fortune in Depression-era America—but as fast as the cash rolled in, George and Kit spent it, on luxury cars, furs, jewelry, lavish seaside vacations, the finest of everything.

Tooling around Fort Worth in their Cadillac roadster, tricked out down to its chrome-spoke wheels, the Kellys seemed like a fairytale couple, the suave, muscular thirty-five-year-old from Memphis, Tennessee, sporting a smartly tailored suit, posing—ironically—as a banker, and the leggy Mississippi-born redhead, eight years his junior, resplendent in her mink coat and diamonds.

George, the son of a well-to-do insurance agent, didn't seem to mind that Kit was a boozy ex-hooker suspected of murdering her previous husband. The Kellys were madly in love and would remain so, just as long as the good times rolled.

Kidnapping Charles Urschel was Kit's idea. He was on a list she had made of wealthy targets. After a dozen bank robberies in three years, the Kellys wanted even bigger paydays. They planned to make a million dollars from kidnappings, then retire to Mexico in style.

At first, the Urschel job went like clockwork. When George returned from picking up the ransom and emptied two hundred fat bundles of cash onto the bed, Kit's eyes lit up. "Mama needs a new pair of shoes," she cooed as she kissed one of the bills. "We pulled it off Kit," George said. "Two hundred thousand bucks, baby."

Kit and Albert Bates, the accomplice who had entered Urschel's sunroom with George, wanted to kill Urschel to silence him, but George told them not to worry: "He could never lead anybody back here. He doesn't even know what state he's in." Just to make sure, George ground the muzzle of a shotgun into the blindfolded oilman's belly and said, "Have you ever seen what one of these can do to a man's face? If you give one bit of information to the cops . . . we'll get to you."

After Urschel's release, Kit celebrated with a shopping spree, but even the purchase of some fabulous jewels couldn't cushion the shock that followed. The news was all over the papers. The Shannon farm had been raided and the owner's twenty-one-year-old son, Kit's stepbrother, had ratted the Kellys out to Gus Jones. The fairytale couple had suddenly become the most wanted criminals in America.

"I don't know whether to kiss you or kill you," Kit snapped, angry at her husband for letting Urschel go. She wrote the prosecutor a bizarre letter threatening him and, in the next breath, offering to turn snitch and set her husband up. George wrote Urschel an equally unhinged letter vowing to leave him "stone dead" and murder his family, signing off with, "See you in hell."

As the Kellys fled, zigzagging their way across the Midwest, dozens of FBI agents fanned out to catch them. The agents ran down hundreds of leads, blanketed the region with wanted posters, even staked out bars and border crossings. And came up empty. At an Indianapolis post office where George had a package waiting for him, he evaded an FBI trap. While checking out a tip outside a Chicago bar, FBI agents overlooked the Kellys, who were inside having a drink. J. Edgar Hoover had hoped the pursuit would showcase the FBI's talents. Instead, the bureau was bungling its first high-profile manhunt.

Finally, after six long weeks, the FBI caught a break. An intercepted telegram led to a twelve-year-old girl who claimed the Kellys were holed up in a small house in Memphis. Among hundreds of bogus tips, this one had a ring of truth.

THE YEGG HUNTERS

Hoover ordered a dawn raid. Leading the way into the lair of America's most infamous gunman was the bureau's Birmingham, Alabama, station chief, Agent William Rorer, a thirty-five-year-old bachelor from Lynchburg, Virginia.

The house was dark and silent as Rorer quietly walked up the steps to the front porch and gingerly tried the door. Strangely enough, it was unlocked. Was he walking into an ambush?

He opened the door and stepped inside. The living room was empty. Two men were asleep in a bedroom. Even in the dim light, Rorer could see that neither was the man he was after. Rorer continued down a hallway. At that moment, a .45 in his hand, Machine Gun Kelly stepped from another bedroom into the shadows behind Rorer.

Rorer's backup, Memphis Police Detective William Raney, aimed his shotgun at Kelly's silhouette and shouted at him. Kelly dropped the .45 and surrendered. Legend would say he threw his hands up and hollered, "Don't shoot, G-men," giving FBI agents their famous nickname. More likely, he just smirked and said, "Okay boys. I've been waiting all night for you." Just before the raid, Kelly had stepped onto the front porch to get the morning paper—and forgot to lock the front door.

Kit woke up mad as a wet hen. She made everyone wait while she put on a sleek black dress with fur epaulets. At least she'd be carted off to jail in style.

Less than three weeks later, October 12, 1933, already tried and convicted for the Urschel kidnapping, the Kellys were sentenced to life in prison. Kit would serve twenty-five years and later die in obscurity in Oklahoma City, at age eighty-one. George would become known in prison as a harmless has-been and get tagged with a new nickname. "Pop Gun Kelly" would serve twenty-one years and die quietly at Leavenworth Prison on his fifty-ninth birthday.

Press accounts touted Detective Raney as the hero of the Kelly raid, while barely mentioning Agent Rorer. (The *New York Times* spelled it "Roper.") Within a few years, the lawman who led the way into Kelly's lair would marry, retire from the FBI, and embark on a successful second career as an executive with a Georgia dairy company. Rorer would pass away at age sixty-nine in 1967.

Gus Jones would serve for forty-three years as a lawman, twenty-eight with the FBI. After finally retiring, he would remain in San Antonio, passing away in his eighties in 1963. His Urschel investigation would go down in FBI history as a brilliant piece of detective work.

Even George Kelly grudgingly agreed. "I should have stayed with what I knew how to do best," he groused in prison. "Robbing banks."

George "Machine Gun" Kelly and his wife Kathryn "Kit" Kelly being sentenced. (ASSOCIATED PRESS.)

THE YEGG HUNTERS

With Machine Gun Kelly out of the way, another gangster took center stage in the War on Crime and would soon become the most famous yegg of all.

Love him or hate him, John Dillinger, the grocer's son from Indiana, was hard to ignore in the early months of 1934. A spree of stickups, shootouts, and jailbreaks kept him constantly in the news. Wanted posters of him were seemingly everywhere.

"I'm not a bad fellow, ladies and gentlemen," he proclaimed during a packed jailhouse press conference, and a lot of Americans agreed. Newsreels of the thirty-year-old with the raffish grin and dimpled chin got so much applause that President Roosevelt took to the airwaves to ask people to stop cheering gangsters. Even a starstruck prosecutor posed with his arm around Dillinger.

It had been ten years since the eighth-grade dropout and deserter from the navy had the book thrown at him for mugging a family friend with a club. In prison, the eager youngster learned all about "cracking a jug"—bank robbery.

Soon after his release, he climbed through the window of a bank in New Carlisle, Ohio after dark and robbed it when it opened the next morning, a quick, easy, and very lucrative $10,000 score. There would be no more penny-ante muggings. John Dillinger had found his calling.

Teaming up with his prison buddies, Dillinger the showman robbed a dozen banks in the Midwest. Typically, entering in his trademark skimmer tilted rakishly to the side, chomping on a stick of gum, he would draw a Tommy gun from a trombone case and grin as he bellowed, "Stick 'em up." After a wisecrack or two—"That's what you get for coming to work on time," he told bank employees during a morning robbery—came his signature move: a leap over the railing in the lobby to get at the cash.

He robbed only the bank, never the customers, and was known for small acts of kindness, draping his coat over a woman hostage who felt cold, that sort of thing, earning him the nickname "Gentlemanly Johnny."

Yet, under the flamboyant charm was a cold-blooded killer. Dillinger had "the coldest eyes you ever saw," a bank teller said. When Dillinger's partners sprang him from a Lima, Ohio, jail, gunning down Sheriff Jess Harber in the process, Dillinger strode casually past him as he lay dying on the floor. Soon, Dillinger himself became a cop killer. Outside an Indiana bank, a police detective shot at him. From twenty feet away, Dillinger turned and raked Detective Patrick O'Malley with a submachine gun, slaying the forty-three-year-old father of three.

Arrested for murder, Dillinger broke out of an "escape-proof" jail in Crown Point, Indiana. To this day no one knows how he came

John Dillinger posing with a prosecutor. (ASSOCIATED PRESS.)

by the fake wooden pistol, crudely carved and blackened with shoe polish, that he used to herd his guards into a cell, before speeding away in the sheriff's car.

The daring escape made headlines, and put Dillinger at the top of the FBI's most wanted list, prompting an order from FDR's attorney general, Homer Cummings: Shoot to kill. John Dillinger was not to be taken alive.

First, "John the Jack Rabbit" had to be caught. Outside his doctor's office in Chicago, Dillinger barreled past a police trap through a hail of gunfire and escaped in a car full of bullet holes without a scratch on him. He got away again after a shootout with local cops and FBI agents at his apartment building in St. Paul, Minnesota.

Hot on his trail was the FBI's Dillinger Squad, twenty agents picked for their daring and marksmanship, men like Charles "Cowboy" Winstead, the gruff, steel-eyed former cattleman and deputy sheriff from Sherman, Texas. Forty-three years old, with a wardrobe consisting of a single threadbare suit and a dirty Stetson, the abrasive Texan was a pariah in a bureau full of eager young preppies. Yet, in his unflattering fitness reports one remark stood out: Winstead "doesn't know the emotion of fear." Besides being fearless, he was a deadeye with his chrome-plated, ivory-handled .38. All in all, a good man to have around for what lay ahead.

The leader of this gung-ho squad was a courtly Southern lawyer. After law school, Melvin Horace Purvis Jr. had wanted to be a diplomat rather than a crime fighter, but the Foreign Service wasn't hiring, so he joined the FBI. The affluent banker's son from Timmonsville, South Carolina, came to work in a chauffeur-driven town car and liked to ride his palomino mare through Chicago's Lincoln Park. Shoot-outs rattled him so much the press dubbed him "Nervous Purvis." Ironically, he was an expert shot, a skill he picked up duck hunting as a boy in South Carolina's tobacco country.

GLORY, GRIT AND GREATNESS

In J. Edgar Hoover's eyes, Purvis was the perfect agent: smart, well-groomed, and dedicated. And short. Small of stature himself, Hoover preferred his top agents that way. "Little Mel" (another press-invented nickname) was just five-foot-seven, 140 pounds. Thirty years old and single, he was a dapper ladies' man whose matinee-idol air dazzled the normally aloof FBI director. "I don't see how the movies could miss a slender, blond-haired, brown-eyed gentleman," Hoover gushed in a letter to Purvis. "All power to the Clark Gable of the service." There were whispers that Hoover, who never married and lived with his mother, had a crush on his protégé.

Within a year of hiring on, Purvis was already a rising star. At twenty-seven he became the bureau's youngest station chief. Three years later he was in charge of the FBI's biggest case ever, the hunt for John Dillinger.

"Get Dillinger for me," Hoover told him, "and the world is yours." What followed, however, was a colossal failure—long, grinding months of dead-end leads, botched surveillances, and missed opportunities. Purvis and his handpicked squad lacked the experience for a large-scale manhunt, and it showed. Dillinger dodged the FBI half a dozen times, most notably at the Little Bohemia Lodge. Finding him got even harder after his features were altered by a plastic surgeon.

But a showdown was coming. When the FBI arrested Dillinger's girlfriend, Billie Frechette, for being his accessory, Dillinger vowed to kill Purvis. On July 22, 1934, outside the Biograph Theater in Chicago, Gentlemanly Johnny would get his chance.

Dillinger liked movies, even gangster flicks, and he was at the theater that night to see *Manhattan Melodrama*, about a racketeer played by Clark Gable. Thanks to the plastic surgery, for the first time since becoming famous Dillinger could safely appear in public. Just to prove it, whenever he passed a cop on the street he looked him in the eye and smiled. When not thumbing his nose at the cops or enjoying Chicago's nightlife, he spent his time planning a train robbery, hopefully such a big score that he could retire in luxury. Despite feeling safe, he still carried a .38 semi-automatic in his pocket.

With him at the Biograph were his new girlfriend, Polly Hamilton, a waitress and part-time hooker, and a second woman, the forty-five-year-old Romanian immigrant who ran the brothel where Polly worked. The woman's name was Ana Cumpanas, which she had anglicized to Anna Sage, and Dillinger had no inkling that she had betrayed him. Seeking reward money and an end to her immigration troubles, she had tipped off the FBI.

Outside, standing near the ticket booth, Melvin Purvis was as nervous as a cat in a snake pit. He was so scared his knees were wobbling. When he saw Dillinger exit the theater, Purvis was supposed to light a cigar. He would have to do it with trembling hands.

For the better part of two hours, while Dillinger had been relaxing in the air-conditioned theater, Purvis and two dozen other lawmen had been sweating in Chicago's searing summer heat. They had the theater surrounded, but Dillinger had escaped from tight spots before. Approaching him on a busy street would be risky. Pedestrians might be in the line of fire. For Purvis, another tragedy like the one at the Little Bohemia Lodge was too awful to contemplate.

Suddenly the agonizing wait was over. A crowd from the packed theater began spilling out of the lobby. In a bizarre twist, Chicago police cars rolled up. The manager had called the cops because he thought the lawmen loitering around the theater were there to rob it. Purvis spotted Dillinger and the two women and lit the cigar, but in all the confusion no one swooped in to help.

The man Purvis had been hunting for months, who had put Purvis on his hit list, walked right at him, close enough for Purvis to reach out and touch him. For a chilling instant their eyes met.

GLORY, GRIT AND GREATNESS

Left: Melvin Purvis and J. Edgar Hoover. (GETTY IMAGES.) *Right*: Anna Sage.

Chilling for Purvis at least. Dillinger's eyes were unsuspecting. Purvis didn't reach for his .45. Pedestrians were in the way. He let Dillinger pass.

Dillinger followed the flow of the crowd down the sidewalk. A few feet behind him, Purvis lit his cigar again. Polly Hamilton was the first to sense the danger. She tugged at Dillinger's shirt, warning him. In one motion he drew his .38 and wheeled around.

At last, three other FBI agents were moving in. Shots rang out. People scattered and a woman screamed in pain. Two bystanders were hit by stray bullets and had minor wounds. Hamilton, Sage, and the FBI agents all escaped injury.

Dillinger, his gun unfired, lay on the pavement. Firing amid the crowd, Agent Charles Winstead, the Texas deadeye, had drilled him in the neck. Dillinger mumbled something that Winstead couldn't make out and then died.

THE YEGG HUNTERS

FBI Agent Charles Winstead

As a crowd gathered, souvenir hunters stepped forward. Men dipped their handkerchiefs, and women the hems of their skirts, into a pool of Dillinger's blood. Disgusted, Purvis walked into an alley and vomited.

Glamorizing the orange skirt she wore to the Biograph, the press would dub Anna Sage "the Woman in Red." She would get a $5,000 reward for setting Dillinger up, but Purvis's efforts to spare her from deportation would fail. She would die in obscurity in Romania.

Although Purvis never drew his gun, the press would dub him "the Man Who Got John Dillinger." The FBI wouldn't publicly acknowledge who fired the fatal shot, but it was well known inside the bureau that it was Winstead. In 1942, facing his third disciplinary transfer, the irascible Texan would tell J. Edgar Hoover, "Go to hell," walk away from an FBI pension, and retire to New Mexico to pursue his first love, cattle ranching.

GLORY, GRIT AND GREATNESS

To his dying day at age eighty-two, Winstead would talk little about the Dillinger shooting, dismissing questions with a standard retort: "Dillinger came out of the theater and died of lead poisoning."

Three months later, in October 1934, four hundred miles from Chicago near the backwoods town of Wellsville, Ohio, Melvin Purvis would face an even deadlier outlaw, an Oklahoma hillbilly with more murders to his name than John Dillinger and Baby Face Nelson combined, someone who had sworn not to be taken alive, a man J. Edgar Hoover had ordered shot on sight.

In the hunt for Charles Arthur Floyd no quarter would be asked and none given.

The husky, five-foot-eight thirty-year-old from Akins, Oklahoma, despised the nickname that had made him famous. Flat-nosed and chubby-cheeked, bearing a passing resemblance to baseball star Babe Ruth, "Pretty Boy" Floyd earned his nickname not because of his face but because of his flashy trappings: the checkered shirts and butterfly bow ties he favored, the gold caps on his front teeth, and the lilac-scented hair tonic he used to slick back his thick brown mane. He once traded five gallons of moonshine for a pearl-handled pistol and carried a gold pocket watch with ten notches in it.

One for each man he had killed.

He shot patrolman Ralph Castner dead following a bank robbery in Ohio, and killed another lawman, Treasury Agent Curtis Burks, during a raid on a Kansas City speakeasy. No one dared call Floyd "Pretty Boy" to his face, especially after two of his rivals who used the nickname to taunt him ended up dead in a roadside ditch.

Between killings, the dirt farmer's son from a strict Baptist family had become perhaps the most prolific bank robber in U.S. history.

THE YEGG HUNTERS

Pretty Boy Floyd in 1929 booking photo.

Floyd and his partners robbed, by his own estimate, sixty banks, mainly in small Oklahoma towns. He was so successful that he hired out as a consultant, helping other gangs plan their robberies, so many all told that the *Daily Oklahoman* called for the mobilization of the National Guard. Hit four times, a bank in Stonewall, Oklahoma, posted a "Notice to Bank Robbers," announcing the bank had almost no cash. Just over the border, the town of Fort Smith, Arkansas, took to guarding its banks with machine gun nests.

Between robberies, "The Phantom of the Ozarks" disappeared into the hills of eastern Oklahoma, where he hid out with the help of family and friends, including local cops. In the impoverished hill country, where banks were widely hated, Floyd was revered as a Robin Hood figure. Lawmen and bank investigators who came looking for him were met with stone-faced silence.

At the height of Floyd's crime spree, the governor coaxed a legendary lawman out of retirement to hunt him down. Forty-six-year-old former sheriff Erv Kelley, who had caught a score of murderers and bank robbers, swore to capture Floyd dead or alive. "It'll be him or me," Kelley boasted, "and I'm not worrying about myself."

At a farm in Bixby, Oklahoma, Kelley and his quarry shot it out. The result was yet another notch on Pretty Boy's gold watch.

Other notches probably stood for the Kansas City Massacre. The evidence pointed to Floyd's being one of the gunmen, earning him a place on the FBI's most wanted list. With so many lawmen after him, he turned fatalistic. "I haven't got a chance except to fight it out," he remarked. "I don't aim to let anybody take me alive."

He moved all the way to Buffalo, New York, throwing his pursuers off the scent for over a year, until he got homesick and set out for Oklahoma. Halfway there, driving in a fog in the predawn darkness, he crashed into a telephone pole on a country road in eastern Ohio. None of the four occupants was badly hurt, but the car was in bad shape.

Floyd sent his two women passengers off on foot to get help, while he and his sidekick and fellow gangster Adam Richetti spread blankets between the trees and settled in with some firearms for what promised to be a long wait.

Later that morning, Wellsville, Ohio, police chief John Fultz was traipsing through the woods outside town, searching for two suspicious strangers who had been spotted in the area. Fultz thought they might be the same two who robbed a bank the previous day in neighboring Tiltonsville.

The forty-six-year-old head of Wellsville's tiny two-man police force was a controversial figure, either a pillar of the community or a scoundrel, depending on whom you asked. Fultz was a churchgoing family man and member of the Elks Club who had withstood accusations ranging from bribery, to consorting with gamblers, to "immoral" conduct with women. Yet, whatever his ethics, one thing was certain. John Fultz, as he was about to prove, was no coward.

That morning he was out of uniform, which was fortunate because as he rounded some tall brush he found himself staring down the barrel of Pretty Boy Floyd's .45. "Don't come another inch," Floyd said, "or I'll pump you." The chief not only ignored the threat, he acted like a man with a death wish. He strode right up to the gun barrel. With the .45 jammed into his belly he made a flip remark, "You wouldn't shoot a working man."

Floyd ordered Fultz to put his hands up. Fultz replied, "I won't do it." Floyd, the killer of ten men, was momentarily stunned by the sheer chutzpah. He stepped aside and let Fultz pass. Rising from a blanket, Adam Richetti pointed his .45 at Fultz and pulled the trigger. Floyd did the same with a Tommy gun. But the chief was even luckier than he was brave. Both guns misfired.

Shouting, "You big yellow son of a bitch," Fultz fired back and missed. Pretty Boy ran off and Fultz chased Richetti through the trees. When a bullet from Fultz whizzed past Richetti's head, he decided he'd had enough. The ruthless gangster threw his hands up and whimpered like a child, "I give up. For God's sake, don't shoot me. Don't kill me. I am done."

Later, Fultz's testimony would help send Richetti, a convicted cop killer, to the gas chamber. For now, the small-town chief was basking in his fifteen minutes of fame. The hunt for Floyd and the daring capture of his sidekick were national news. Reporters flocked to the scene. Fultz posed for pictures beside a shackled Richetti. A reporter asked if Fultz feared revenge from Floyd's gang. "If the gang is as yellow as Floyd and Richetti," growled the hero of the hour, "I wouldn't be afraid of a carload of them."

Meanwhile, Pretty Boy Floyd had highjacked a car and was a mere hundred yards from the main highway out of town when he hit a police roadblock. Armed with his .45, he ditched the car and disappeared into the woods.

Two days later, with Floyd still at large, Melvin Purvis and a squad of eight FBI agents and local lawmen were slogging through the muddy fields a few miles from Wellsville, checking every farmhouse, barn and chicken coop. Two hundred cops had descended on the region to search for Floyd, and he had eluded them all.

For Purvis's squad, the search had been a major ordeal. Purvis hadn't slept in days and he and his men had been subsisting on apples and pears they'd picked from trees. A crazed sheep had chased Purvis away from one farm, and in a heart-stopping moment he had nearly traded shots with a fellow agent while they were both snooping around the same barn. Before quitting for the day, Purvis and his men drove to one last farm, owned by a woman named Ellen Conkle.

At that moment Floyd was in her driveway. She had just fed the pitiful looking stranger who happened upon her farm after hiding in the woods all weekend. Her brother was about to give Floyd a lift to a bus station.

As Floyd concealed himself behind a corn crib, Purvis spotted him and yelled at him to surrender. Instead, Floyd bolted toward a shallow rise, zigzagging to dodge the gunfire he knew was coming. Once he cleared the rise he could disappear into the woods again. He ignored commands to halt. Finally, Purvis shouted, "Let him have it."

At seventy yards, Tommy guns, rifles, and shotguns blasted away. Mrs. Conkle's ears would ring for a week. Dozens of rounds whistled through the apple trees. Just below the crest of the ridge Pretty Boy collapsed, mortally wounded.

As the cops approached, he tried to aim his .45 at them, but his shattered arm was useless. When an agent asked him about the

Kansas City Massacre, Floyd sneered, "I ain't telling you nothing. Fuck you." Within a few minutes he was dead.

Six days later, in Oklahoma's largest funeral ever, twenty thousand people bid Charles Floyd farewell. The minister lauded him as a working-class hero. Folk singer Woody Guthrie would lionize him in a song: "The Ballad of Pretty Boy Floyd." A biographer would note Floyd's "courage" and "compassion."

J. Edgar Hoover summed up another school of thought. Pretty Boy Floyd, he told reporters, was nothing but "a yellow rat who needed extermination."

His mother would always say her boy Lester was the most beautiful of her seven children. He had the golden locks and round cheeks of a cherub, to go with gleaming white teeth and blue eyes. "He had the bluest eyes I've ever seen," one of his victims would remember. "I always wondered how someone so innocent looking could be robbing us." Another victim trying to describe him to police would give him his nickname.

George "Baby Face" Nelson (True name: Lester Gillis), the son of middle-class Belgian immigrants, started ditching school in the fourth grade. By eleven he was snatching money from cash registers and stealing cars. By twelve, he was on his way to reform school for what he considered a prank, shooting into a group of boys, one of whom was hit in the jaw by a ricochet.

In and out of trouble, he was warned that when he reached adulthood he would face stiffer penalties. He didn't listen. The twenty-one-year-old and his gang launched a spree of home-invasion robberies in his hometown of Chicago, bursting into the mansions of the wealthy, binding and gagging the occupants at gunpoint, and making off with their jewelry.

Next, he started robbing banks, using a potent new weapon. The first time he got his hands on a Tommy gun, he was as tickled as a kid on Christmas morning. His calling card became a hail of lead delivered on his way out of town. After his robberies, he blasted away at storefronts, cars, plate-glass windows, even a theater marquee, sending glass and chunks of cement flying and pedestrians diving for cover.

He was suspected of murdering two civilians, a motorist gunned down in a road-rage incident and a government witness whose body was never found, but above all Baby Face delighted in shooting cops. A witness heard him "chirp with joy" as he shot Officer Hale Keith at a bank robbery in Sioux Falls, South Dakota. Hit with four slugs from Nelson's Tommy gun, Keith would miraculously recover. At a robbery in Mason City, Iowa, Nelson fired at a bystander, hitting him in the leg, then walked up to the bleeding man and sneered, "You stupid son of a bitch. I thought you were a cop." On a backroad outside Chicago, Nelson blazed away at two Illinois state troopers sitting in their squad car, wounding both.

Baby Face Nelson. (SHUTTERSTOCK.)

THE YEGG HUNTERS

His tally of six lawmen shot and one killed put him on the front page. With John Dillinger and Pretty Boy Floyd dead, Baby Face Nelson, age twenty-five, took his turn as the FBI's Public Enemy Number One.

He's a runty little murderer. Wherever the trail leads, our men will follow until Baby Face is behind bars or beneath the dirt where he belongs.

—FBI AGENT SAM COWLEY TO A REPORTER

An FBI administrator, a desk man rather than a street cop, Agent Samuel Parkinson Cowley seemed ill-suited to square off with a killer as dangerous as Baby Face Nelson.

Cowley even looked unimposing. With a rounded chin, thin mouth, pug nose, eyes set too close together, and slicked down hair parted conservatively on the side, he had a plain, earnest face, neither homely nor handsome. Thirty-four years old, five feet nine inches tall and slightly pudgy, he could have passed for a middle-aged insurance agent, or, if he had worn expensive suits, a banker.

He may not have been a sharp dresser, said an early fitness report, but his mind was razor sharp. And he drove himself relentlessly, working from early in the morning to late at night. When his second son was born, the new father was too busy to go to the hospital or even think up a name for the child, so his wife Lavon fell back on Samuel Cowley Jr.

Within five years of joining the bureau fresh out of law school, Cowley was named the bureau's chief of investigations, a job that rarely took him away from his desk. Seeing no reason to hone his firearm skills, he didn't bother to qualify at the bureau's shooting range.

When the Dillinger manhunt stalled, J. Edgar Hoover put Cowley in charge, moving Melvin Purvis to second-in-command. Cowley didn't mind that the press gave Purvis credit for getting Dillinger. The pious Mormon preferred to avoid the spotlight.

Next, under Cowley's leadership, the FBI went after Baby Face Nelson with a vengeance, payback for the murder of Agent Carter Baum. After spending a mistake-filled year chasing Machine Gun Kelly and John Dillinger, this time the bureau got it right, casting a tight net from Chicago to California, keeping Nelson on the run, forcing him to spend his nights in roadside lodges.

Outside Reno, Nevada, Nelson got word that FBI agents were poking around town. Spooked, he fled east, where, in the sticks north of Chicago, he would cross paths with Cowley and a slender young FBI agent with a killer instinct.

Even in a bureau full of fresh-faced recruits, Agent Herman Hollis's youthful looks stood out. Tall and lanky, with a pencil neck, thin face, and smooth complexion, the thirty-one-year-old looked like he belonged in high school.

A devout Catholic who carried a rosary in his pocket, Hollis also had a grittier side. He had won a medal for his marksmanship and was one of an elite band of agents earmarked for dangerous work. The night John Dillinger was shot, Hollis stood shoulder-to-shoulder with Charles Winstead, part of a three-man team handpicked to take the fugitive down outside the Biograph theater.

But nothing could prepare Hollis for the gun battle he and Cowley were headed into—the three savage minutes named for a sleepy Illinois town.

Working nearby, Harold Kramer heard "The Battle of Barrington" before he saw it. What sounded at first like a car backfiring had erupted into the roar of shotgun blasts and blazing machine guns. "All hell broke loose," Kramer would recall. "Bullets were

FBI Agents Sam Cowley (*Left*) and Herman Hollis (*Right*).

flying everywhere. It was like being in a bunker with a war going on right outside."

The gunfire was coming from the place just off the highway where Baby Face Nelson and Johnny Chase's Ford and Hollis and Cowley's FBI car had come to a halt, thirty feet from each other.

Crouched behind the cars, shell casings piling up at their feet, the four men were locked in a fight to the death. Witnesses looking on in horror fled indoors as stray bullets hit the gas stations across the highway. A terrified driver swerved off the road and flattened himself in the mud. "The shots were terrific," motorist Paul Sherman would remember. "They were buzzing and cracking all around my ears."

The agents drew first blood. His liver pierced by a round from Cowley's Tommy gun, Baby Face Nelson was gripped by a psychotic rage. He moved into the open and walked toward the agents' car, firing as he went. Cowley, supposedly no marksman, managed to pump six more rounds into Nelson's chest and belly. Stunningly, like a creature from a horror movie, Nelson kept advancing. Hollis fired his shotgun into Nelson's legs, felling him. Nelson got up and kept firing.

Cowley and Hollis were hit. With two bullets in his torso, Cowley fell and rolled into a ditch. He rose and tried to fire, but his gun was empty. Wounded, Hollis staggered across the highway toward the safety of a telephone pole. Before he got there, a bullet blew out the back of his skull.

As the smoke cleared, Nelson and Johnny Chase, who was unharmed, abandoned their disabled Ford and fled in the FBI car.

"Was Hollis hurt?" Cowley murmured to a police officer. "Look after him before me." But Herman Hollis was beyond hope. He never regained consciousness. Cowley was rushed to the hospital. "It was Nelson and Chase," he told Melvin Purvis before being wheeled into surgery. "I emptied a Tommy at them [but Nelson] wouldn't go down." With Lavon and his two small sons at his side, Sam Cowley hung on into the night and then died.

The two agents would never know that they hadn't died in vain. The next day, wrapped in a blanket, Baby Face Nelson's naked, bullet-riddled corpse was found in a cemetery where it had been dumped. Soon, he would be beneath the sod where Sam Cowley had vowed to put him.

Arrested a month later, Chase would spend thirty years in prison for murder, before being paroled over the vehement objection of J. Edgar Hoover. Chase would die in 1973, age seventy-one.

The War on Crime officially ended in mid-1936, with the arrest of bank robber and kidnapper Alvin "Creepy" Karpis in New Orleans, the fourth Public Enemy Number One the FBI had killed or captured in less than two years, earning the bureau wide acclaim. After a rocky start, the war had turned the little-known FBI and its director into household names, riding a wave of popularity that would last decades. Hoover would remain FBI director for nearly fifty years until his death in 1973 at age 77.

Melvin Purvis would survive the War on Crime, but his FBI career wouldn't. Disillusioned with Purvis and jealous of the press coverage he had received, Hoover turned on his former protégé and drummed him out of the bureau. Purvis went on to a varied career as, among other things, a small-town newspaper publisher, a colonel in the army, and special counsel to a U.S. Senate committee.

Sadly, plagued by bad health, hooked on alcohol and pain pills, perhaps tormented by a violent past, he would die in 1960 at age fifty-six of a self-inflicted gunshot wound, an apparent suicide. His gravestone contains a Latin inscription that translates as "I was often afraid, but I never ran."

The Kansas City Massacre, the shoot-out that triggered the war, would never be officially solved. By the late 1930s the prime suspects, including Pretty Boy Floyd and Adam Richetti, were all dead and the FBI's case was closed. No one is sure to this day whether the gunmen were out to murder the prisoner, Frank Nash, or free him.

Half a dozen lawmen gave their lives fighting the War on Crime: FBI Agents Carter Baum, Samuel Cowley and Herman Hollis, retired sheriff Erv Kelley, Officer Ralph Castner, and Detective Patrick O'Malley.

These six and a handful of their fellow lawmen are the war's forgotten heroes. Thanks to them, gangs of vicious thieves no longer terrorized the Midwest. Law and order had finally been restored.

America's heartland was safe again.

CHAPTER 9

I AIN'T NO MUSEUM PIECE

John Basilone

Outdoors on a sunny, late summer day in 1943, on a makeshift stage near the town of Raritan, New Jersey, a beautiful film actress named Louise Allbritton, playing to the newsreel cameras, kissed the guest of honor. But the cameras didn't catch it, so she kissed him again, as a huge crowd roared its approval and the object of her affection, a tough-as-nails Marine, turned beet red.

That was the day, someone would recall, that "the world came to Raritan." Thirty thousand people had lined a parade route stretching all the way to neighboring Somerville. A dozen brass bands thundered by, along with drill teams, color guards, drum-and-bugle corps, and row after uniformed row of soldiers, police officers, firefighters, American Legion veterans, Boy Scouts, Girl Scouts, air raid wardens, Red Cross workers, and nurse's aides. Women Marines marched in heels. The Elks and Knights of Columbus joined in. There was even a troop of French sailors.

GLORY, GRIT AND GREATNESS

Every group in the endless procession received polite applause until at last the crowd went nuts. Seated in a convertible between his parents Salvatore and Theodora, the very same guest of honor, the hometown boy everyone had come to see, had rolled into view. All along the route people ran into the street to greet him. Everyone wanted to shake his hand. Slap him on the back. Bask in the glory of America's most celebrated fighting man.

Afterward, at a war-bond rally, in front of a sea of people and a grandstand full of VIPs, he stepped to the microphone and wrestled his way through a few lines about "backing the attack." Though no orator, he was a publicist's dream, a humble working stiff turned dashing war hero, a strapping twenty-six-year-old with wavy black hair, bedroom eyes, and dimples when he smiled, a broad-shouldered Adonis in a smartly tapered Marine Corps tunic.

No wonder he would help sell over a million dollars in war bonds that day, no small sum in 1943. No wonder that, of the thousands of letters he'd received, no small number had been marriage proposals. "I always wanted to marry a hero," gushed one smitten fan.

But what was most impressive, what set him apart from the legions of other brave young men fighting the Germans and Japanese overseas, was the star-shaped hunk of brass depicting the Roman goddess of war that hung from a powder-blue ribbon around his neck.

Of the fifteen million Americans who fought in World War II, only about two hundred would win the country's highest award for valor, the Congressional Medal of Honor, and live to tell about it.

Among the first was Marine Sergeant John Basilone (*BĂZ-ĭ-lōn*), who came home to a hero's welcome but soon grew tired of the limelight, who would leave everything behind—the glitter, the glamour, the starlets, the cheering crowds, even the woman he adored—to go back to the war in the Pacific.

Back to the horror of a night that still haunted him.

I AIN'T NO MUSEUM PIECE

Turning toward their homeland four thousand miles away, thirty five hundred Japanese soldiers, the crack 29th Infantry Regiment, chant an ancient oath:

Swollen corpses drifting in the sea depths,
Corpses rotting in the mountain grass,
We shall die; we shall die for the emperor;
We shall never look back.

Near midnight, bayonets fixed, the regiment attacks, storming out of the jungle toward seven hundred United States Marines.

It is October 24, 1942, on Guadalcanal Island in the Southwest Pacific's Solomon chain, at the southern edge of Japan's newly conquered empire. After more than two months of hard fighting, the Marines cling to a toehold, an eight-mile perimeter, bulging down from the island's north shore. Tonight, the Japanese plan to smash through the thinly held line at its weakest point, pour their reserves through the breach, and drive the First Marine Division into the sea.

The immediate target is an airstrip named for a Marine pilot killed in action—Henderson Field. The Japanese commander, General Masao Maruyama, has vowed to crush the Marines around the airstrip "in one blow." Even as the battle rages, he is so sure of victory that he broadcasts the code word that Henderson Field has fallen—*Banzai*.

He has reason to feel confident. Japan's soldiers are among the most fearsome in the world, the conquerors of Singapore, the Philippines, and the Dutch East Indies, men as ruthless as they are brave, who fight to the death and behead their prisoners. The 29th,

GLORY, GRIT AND GREATNESS

Maruyama's spearhead, is the cream of the cream, a force of battle-hardened veterans trained to a razor's edge.

Blocking their way, little more than a skeleton force stretched along a mile and a half of the American line, is a lone battalion of Marines, the First Battalion of the 7th Marine Regiment, commanded by legendary Lieutenant Colonel Lewis "Chesty" Puller. Earlier in the day, while checking his defenses, Puller came across a sergeant digging a machine-gun nest at the center of the line. "Son," Puller deadpanned, "if you dig that hole any deeper, I'll have to charge you with desertion." John Basilone grinned. His foxholes are indeed deep, allowing him and the dozen-odd men under him to fire their four machine guns while standing up.

Basilone loves three things above all: good hootch, headstrong women, and the water-cooled Browning .30-caliber machine gun. Though heavy (ninety pounds with its tripod) and prone to jam,

Sergeant John Basilone wearing the Medal of Honor. (NATIONAL ARCHIVES.)

I AIN'T NO MUSEUM PIECE

when properly handled, a Browning can mow men down like a scythe cutting wheat, and Sergeant Basilone is just the man to do it. He has trained on the weapon for years. He can even break it down and reassemble it with lightning speed blindfolded—a skill that is about to save his life.

The October night is nearly pitch-black, the jungle beyond the perimeter barely visible in the moonlight trickling through the clouds. Snipping noises echo through the darkness. The Japanese are cutting barbed wire.

A red flare goes up. More than a hundred Japanese soldiers, screaming "Marine you die!" rush Basilone's foxholes. Muzzle flashes from four Brownings light up the night. White-hot shell casings—and enemy corpses—pile up, as belt after belt of ammo rattle through the guns.

The Japanese fall back and attack again. Enemy soldiers spread-eagle themselves on the barbed wire, as their comrades swarm over them. More bodies pile up. The Japanese lob grenades and mortar shells, trying to blast Basilone and his men out of their holes. Basilone's squad holds on, hurling back one suicide charge after another. The machine-gun barrels get so hot their water jackets are running dry. Basilone tells his men, "Piss in 'em!" Twice during the night, dashing past enemy snipers who have infiltrated into the trees, Basilone sets out for supplies. He lumbers back through the jungle dodging bullets as he lugs water, spare barrels, and masses of bandoliers draped over his shoulders.

The two machine guns on his right flank are knocked out. Basilone picks up a ninety-pound Browning, orders two men to follow him, and heads that way, firing from the hip, scorching his arm on the searing barrel, as he and the others annihilate a squad of Japanese. Under fire, working frantically in the dark, he manages to fix one of the guns. Flat on the ground, he rolls back and forth between the two Brownings, pouring lead into yet another assault. Someone shouts, "Look out!" Enemy soldiers have slipped past the blazing guns and are attacking from the rear. Basilone wheels around and shoots them down with his .45.

By now, the fight has lasted six hours. The ammo is all but gone, and Basilone's command is down to him and two others. In the nick of time, reinforcements are fed into the line. At last, the Japanese withdraw for good.

The Marine perimeter has held. The Stars and Stripes still flies over Henderson Field. A thousand Japanese lie dead, including scores of them covering the blood-soaked ground defended by Basilone and his men. Maruyama's vaunted 29th has been shattered. Basilone's Medal of Honor citation will credit him with "contributing in large measure to the virtual annihilation of a Japanese regiment."

But he isn't thinking about a medal. The first thing he does after the battle, without food or rest, is stop by a hospital tent to check on his men. "He was barefooted," a wounded Marine will remember, "and his eyes were red as fire. His face was dirty black from gunfire and lack of sleep. His shirt sleeves were rolled up to his shoulders. He had a .45 tucked into the waistband of his trousers. He'd just dropped by to see how I was making out. Me and the others in the section. I will never forget him."

The kid everyone called "Bazzy" had started out as a happy-go-lucky goof-off, best known for being "the most talkative boy in class" and for downing the most ice-cream sodas at the local soda fountain. His favorite hobby, said his eighth-grade yearbook, was gum chewing.

Unable to sit still—a textbook ADHD case before the term existed—he got his knuckles rapped often by the nuns at

Top: Marines on patrol on Guadalcanal. (NATIONAL ARCHIVES.) *Bottom*: Unidentified Marine firing a water-cooled Browning .30 caliber machine gun.

St. Bernard's Parochial School. Nearly as often, he cut class to skinny dip at "Bare-Ass Beach," his favorite swimming hole. He and his fellow scamps threw popcorn at movie screens, rotten tomatoes at girls, and filched apples from an orchard, the last of which earned a hiding from his mother. "I smacked him good," she later told reporters.

Yet no one could deny the kid had pluck. At age seven, with more brass than brains, he snuck into a neighbor's corral to try his hand at bullfighting. The bull took pity on him and merely knocked him into the mud.

By age ten, the youngster was proving himself with his fists. Fighting skills came in handy in blue-collar Raritan, home to tough Italian, Polish, and Irish kids. Bazzy wasn't one to pick fights, but, especially if a friend was being bullied, he didn't back down either. He could throw a punch and take one. He didn't mind occasional black eyes. They impressed the girls. He dreamed of being a champion boxer. Or an opera singer.

Like millions of native Italians, his father had dreamed of starting a new life in America. Sal Basilone emigrated from Naples at age nineteen and scrimped and saved for years until he could afford to open a tailor shop. He became a staunch American patriot, bristling whenever someone badmouthed his adopted country. He and his vivacious wife Dora, a first-generation Italian American, would raise their ten children in a very crowded, very lively, three-bedroom one-bath household, and send three sons off to war.

Against his parents' wishes, Bazzy dropped out of school after the eighth grade and bummed around town for a couple of years, caddying part-time at the local country club and delivering laundry until he got fired for sleeping on the job. "I feel like a misfit," he confided to his priest.

For the restless seventeen-year-old, getting fired was a badly needed kick in the pants. One night at the dinner table, he announced he was joining the Army. "You're crazy," his father said. "You're only a kid." On second thought, Sal relented. Maybe the Army would make a man out of his wayward son. And indeed it would.

With help from a fiery Filipina.

In the Army, Private John Basilone would learn how to march in step, handle a rifle, and field strip a machine gun. And how to revel.

"He pulled some pretty wild liberties," a buddy chortled seventy years later with a gleam in his eye. "John was a live wire."

He spent most of his hitch in Manila, the capital of the Philippines, where his off-duty haunts were the bars and brothels along the city's waterfront. He drank "blockbusters," a concoction of scotch, rye whiskey, bourbon, and gin. "You could start sipping in the morning," his buddy said, "and you couldn't find your ass with both hands by dinner."

Basilone went wild over the women, lithe, dark-eyed beauties who reminded him of the Italian girls back home. Sometimes in Manila he paid for sex, other times the young Victor Mature lookalike with the winning smile didn't have to. Amid all the reveling, Basilone began a torrid affair with a "bar girl."

Little is known of her except her name, Lolita. The couple ran a bar together, a common sideline for servicemen overseas, and she became his guide to "The Pearl of the Orient," as Manila was known, taking him through the exotic world along the Pasig River, past huts built on stilts over the water, to Intramuros, the walled city of the Spanish conquistadors, and to the teeming markets of the world's oldest Chinatown. She introduced him to her parents. He wrote home about her. Eventually, the young lovers talked about getting married.

GLORY, GRIT AND GREATNESS

The fairy tale ended when it came time for him to ship out. When she found out he meant to leave without her, she went searching for him with a forlorn heart. And a machete. Unable to find him, she hacked his seabag to pieces.

He left town one step ahead of the knife, with a boxing title and a new nickname that would stick with him. "Bazzy" was now "Manila John," the undefeated light-heavyweight champion of the U.S. military base. None other than General Douglas MacArthur himself had dropped by the ring to wish him luck. Basilone's showdown with the U.S. Navy's champ had been a big event. Before a raucous crowd, Manila John and "Sailor Burt" had traded knockdowns until Basilone KO'd him.

Boxing laurels. Alluring women. A lot of booze and sex. In many ways Manila had been good duty. Basilone's physique had filled out, and, when not partying, he'd thrived on the training and discipline.

Private John Basilone, U.S. Army

He'd gotten a cool tattoo, a dagger that ran down his upper arm with the caption *Death Before Dishonor*, and there were even high jinks, like the time he and some buddies busted a fellow soldier out of the local drunk tank by crashing a truck through the jail wall.

But there were also godawful heat and swarming mosquitos, and in the peacetime Army, long hours of boredom. And no Italian home cooking. When his enlistment was up, the twenty-one-year-old mustered out and headed home. He walked in on his mom unannounced. Dora had barely dried her tears of joy before her son was wolfing down his first plate of spaghetti in three years.

He found a decent enough job working as an installer for a gas company, but before long felt restless again. He decided to reenlist, but not in the Army. War was on the horizon, and he wanted to be in the thick of the action with the toughest outfit around, the United States Marines. This time Sal willingly gave his blessing.

"Thanks Pop," said his son. "Someday, I'll make you proud of me."

That day came in June 1943 via a letter postmarked Melbourne, Australia:

> I am very happy for the other day I received the Congressional Medal of Honor, the highest decoration you can receive in the armed forces. Tell pop his son is still tough.

Melbourne had rolled out the red carpet for the Marines who'd fought on Guadalcanal. They were hailed as the saviors of Australia, who had driven the Japanese back from the gates of the Land Down Under. There were brass bands and cheering crowds, free drinks and sumptuous meals at plush hotels. And, best of all, flocks of women.

"We had strawberries and cream," said Sergeant Richard Greer, Basilone's drinking buddy. "All the Australian guys were out fighting the war. The girls were deprived for so long. We tried to remedy that situation."

Basilone made no bones about his debauchery: "One thing I learned about nearly dying is that it gave me a hell of an appetite for everything. . . . I ate, drank and screwed like a wild pig, and didn't feel bad for one second about any of it."

Yet, no amount of carousing could erase the terror of that night in the jungle. PTSD plagued the hell-for-leather hero. Staring down at his liquor glass, he saw enemy snipers, heard the pinging of their bullets. He told his sister of nightmares that left him trembling and drenched.

He refused to believe the rumors that he was being awarded the Medal of Honor. He didn't think he deserved it. Guadalcanal had claimed thousands of American lives. Why should he be singled out?

At the medal ceremony, as his citation was read over a loudspeaker, "a thousand thoughts" raced through his mind. What would his parents think? Would generals have to salute him? Recalling his dead buddies, he beat back tears. He would always insist "pieces" of the medal belonged to the men who fought beside him. Though fiercely proud of it, he would rarely wear it. The memories were too painful.

Soon after the ceremony, he was ordered back to the States. On leaving his platoon, he was said to have wept.

He arrived back home to popping flashbulbs and hordes of reporters. Starting with a press release about the medal ceremony, Basilone's story had snowballed. Sergeant John Basilone had become "Manila John the Jap Killer," a superhero in dungarees. He got an inkling of the hype when he was asked to autograph a comic book he was featured in.

I AIN'T NO MUSEUM PIECE

Medal of Honor ceremony, Camp Balcombe, Australia, May 21, 1943. Four U.S. Marines are awarded the medal for their service on Guadalcanal: (*left to right*) Major General Alexander Vandegrift; Colonel Merritt Edson; Lieutenant Mitchell Paige; Sergeant John Basilone.

"Hell, Johnny," his brother told him. "You ain't seen or heard anything yet. The whole country is crazy about you. You're a hero and famous all over the world. Everybody from President Roosevelt on down is talking about you."

"Jersey Marine Gets Honor Medal" was a front-page headline in the *New York Times*. Life magazine ran a photo spread. *Collier*'s magazine put a rendering of Basilone—mud-spattered, draped with bandoliers, cradling a machine gun under bulging biceps—on its cover.

He was honored with parades and stacks of fan mail, and welcomed into the rarified world of New York City's toniest nightclubs, including the hottest spot in town, Toots Shor's, where Basilone drank for free and Toots introduced him to the city's glitterati, the entertainers, sports stars, and glamour girls who gathered at the club to see and be seen.

GLORY, GRIT AND GREATNESS

Basilone could take hero worship or leave it, except when it came to children. Kids would gather outside his house to get a glimpse of their idol, and he delighted in having them swarm all over him.

The Navy brass turned the beloved hero into a reluctant cash cow, an awkward pitchman hawking war bonds on radio and in newsreels. He was ill at ease on the air even when working from a script.

Most nerve-racking were the roomfuls of reporters firing questions. *Hey Johnny, were ya scared? How'd it feel when ya got the medal? Do you like being called Manila John the Jap Killer?* Dripping sweat as he faced the Washington press corps, he turned to his handler and muttered, "This is worse than fighting the Japs."

He was sent on a war-bond tour that featured an array of Hollywood stars—John Garfield, Eddie Bracken, and others—all of them

This rendering of John Basilone during the Battle for Henderson Field was the cover of *Collier's* magazine on June 24, 1944. (PAINTING BY C.C. BEALL; COURTESY OF THE NATIONAL MUSEUM OF THE MARINE CORPS, TRIANGLE, VIRGINIA.)

outshone by the Jersey Marine. In Madison Square Garden nineteen thousand people gave Basilone a standing ovation. At munitions plants, workers crowded around to say, "Attaboy." No one knows how many war bonds he sold, but the dollar amount ran well into the millions.

Bringing more than a touch of class to the tour was a dazzling honey-blond actress named Virginia Grey. She introduced Basilone to the crowd at a rally and then turned and slipped him a come-hither look. After a grueling day on the road, the two of them would steal away to a secluded booth in the hotel bar to unwind over drinks. She was witty, urbane, and drop-dead gorgeous. The former "most talkative boy in class" had his own streak of charm and was nearly as good-looking. Sparks flew, but whether the romance remained platonic is anyone's guess. After the tour was over the couple drifted apart.

Adored by women, revered by men, cheered and feted wherever he went, Basilone was living the kind of life people dream of, and he hated it. He began drinking more than ever, much more, draining a bottle of hard liquor the way most people knock back a beer. Life on a pedestal had worn thin. The gung-ho warrior wasn't cut out to be a star. He wanted to be back behind a Browning, shoulder-to-shoulder with his fellow Marines.

"I ain't no museum piece," he told his superiors. "I belong back with my outfit." He was offered a promotion to lieutenant and a cushy stateside gig, both of which he turned down. His requests for combat duty were shelved until finally at a black-tie dinner he ran into his commanding officer on Guadalcanal, Lieutenant General Alexander Vandegrift.

There were no cameras, no script. Just a heartfelt plea, one mud Marine to another. It worked. Somehow the red tape fell away.

GLORY, GRIT AND GREATNESS

Virginia Grey

A few months later, in the spring of 1944, John Basilone was as happy as he'd ever been. Stationed at Camp Pendleton in sunny Oceanside, California, the cameras, the crowds, and the bond tour all thankfully behind him, he was doing his dream job, schooling a platoon of fifty baby-faced Marines, whom he would soon lead into battle, in the fine points of machine guns.

He was now a gunnery sergeant, a promotion that came with yet another nickname. To his young platoon mates, "Gunny" Basilone was equal parts drinking buddy, father figure, and drill instructor, someone who could pull wild liberties alongside his men, drink them under the table, and crack the whip the next morning, and through it all, command undying respect and admiration. The kids in the platoon worshipped him.

I AIN'T NO MUSEUM PIECE

One of them would recall Basilone's "simplicity, his cheerfulness . . . the charm and easy grace with which he carried his honors [which] gave us not only confidence, but pride. We were 'Basilone's boys' and were envied for it throughout the division."

Far less impressed was a woman sergeant who worked as a cook, a thirty-one-year-old Italian American, slender and sassy, with a smile out of a Pepsodent ad, a smile so bright you could spot it a mile away—or so said John Basilone.

The first time he saw it he was hooked. He looked her in the eye from across the mess hall and nodded, an invitation any number of women would have killed for.

But Sergeant Lena Riggi was no starry-eyed groupy. She was smart and tough, capable of cutting any man off at the knees, even a Medal-of-Honor-winning hunk. "She shrugged him off," said her roommate. "That's John Basilone. The John Basilone," Lena's friends told her. "So what?" she sniffed.

His Plan B was to invite her along on group outings, a bunch of Marines on liberty taking the train up to Los Angeles. One night in the middle of an L.A. street for all to see, he kissed her and shouted, "I'm going to marry this girl!"

This time the ice queen melted. His gentle charm and good-natured swagger had won her over. A sergeant's pay couldn't buy many glamorous evenings, but that hardly mattered to a couple so deeply in love. Their dates were movie nights followed by sodas at the PX, or talking for hours during long walks around the base with tanks and trucks for scenery. She spoke of growing up on a farm in Oregon in a large Italian family like his. He told her of his dream of having ten children, like his parents did, but in a big ten-bedroom house.

With John about to ship out, he and Lena, like countless other wartime lovers, became a couple in a hurry. She found a priest

Mr. and Mrs. John Basilone.

willing to marry them on short notice, and she and John tied the knot in a small but joyous ceremony in an Oceanside chapel, with a few Marine friends and lots of rice and warm wishes. The honeymoon was a quick getaway to Los Angeles on a seventy-two-hour pass, three blissful days spent painting the town at night and lounging in bed all morning and afternoon.

Two weeks later, his orders came down. There was only time for an all-too-rushed goodbye in the mess hall where the newlyweds had met. They talked about their plans after the war, what they would name their first son and first daughter. John promised to return. Somehow Lena managed not to cry.

She gave him the cross she wore, praying it would keep him safe as he headed into "a nightmare in hell."

I AIN'T NO MUSEUM PIECE

That was how *Time* magazine's Robert Sherrod described Iwo Jima—"Sulfur Island"—eight-square-miles of barren volcanic rock in the western Pacific, an island whose most noticeable feature, other than five-hundred-foot Mount Suribachi, would soon be vast rows of white wooden crosses marked "USMC."

In February 1945, Iwo Jima was the U.S. Navy's next stepping stone en route to the Japanese mainland, seven hundred miles away. The Navy had pounded the island with thousands of tons of high explosives. The brutal work ahead would be left, as usual, to the Marines.

To everyone's surprise, four waves of leathernecks hit the beach unopposed. "Where are the Japs?" wondered Private Chuck Tatum. They were nowhere—and everywhere. Twenty thousand of Dai Nippon's finest, with orders to die for their emperor, lay in wait, hidden in bombproof caves and blockhouses bristling with big guns.

The Japanese had set a trap, holding their fire until the beachhead was packed with men and equipment. Tatum heard the "banshee wail" of a single shell, and in the next instant all hell broke loose, shattering the eerie calm with deafening blasts, soon followed by blinding smoke, screaming men, and mangled bodies.

Pinned down, unable to dig foxholes in the loose volcanic sand, thousands of Marines hugged the ground under a torrent of lead and steel. Victims of direct hits literally disintegrated, vanishing with barely a trace. Other men were tossed in the air like rag dolls. Severed limbs littered the beach. The shoreline became so clogged with wrecked vehicles and human carnage that boats bringing in fresh troops had trouble finding space to land.

Amid the slaughter, Tatum spotted "a lone Marine walking back and forth on the shore" as though in a training exercise.

GLORY, GRIT AND GREATNESS

D-Day on Iwo Jima. Pinned down, Marines hug the beach below Mt. Suribachi.

It was John Basilone, slapping helmets, kicking butts, ordering men to get up and follow him inland.

He led a patchwork twenty-man squad in a textbook assault on a blockhouse. One man blew open the steel doors with a satchel charge, another flushed the enemy with a flame thrower, and Basilone stood on the roof with a Browning on his hip, mowing down the survivors as they fled.

From there, crouching as they ran, he and his men raced ahead to a giant shell hole at the edge of a main objective, an airstrip half a mile behind the beachhead. When Tatum looked around, his heart sank. The squad was alone, far forward of any other Marines, with shells falling all around. The tiny outpost badly needed reinforcements. Basilone shouted, "Hold this ground come hell or high water. I'll go back for more men," and took off.

On Guadalcanal he'd made a mad dash through sniper fire. Now, there were snipers, machine guns, mortar fire "coming down

I AIN'T NO MUSEUM PIECE

like rain" (Tatum's words), even land mines. Under heavy fire, Basilone stopped along the way to help direct the fire of some Sherman tanks—in the middle of a minefield.

Back at the shell hole, Tatum stuck his head up. This time what he saw made him want to cheer. Here came Gunny Basilone to the rescue, leading a squad of reinforcements just as he promised. He'd made it to the beachhead and back without a scratch. His legendary run of luck had held. He was as invincible as ever.

Until he wasn't. A stone's throw from the refuge of the shell hole, a mortar shell exploded at his feet, killing him.

In the blink of an eye, at 10:45 a.m. on February 19, 1945, two hours into the Battle of Iwo Jima, John Basilone, age twenty-eight, America's larger-than-life superhero, was gone, along with his dreams. He would never hold Lena again. Never have that big house full of kids.

"Sergeant Basilone must have known in his heart that his luck wouldn't last forever," wrote the *New York Times*. "Yet he chose to return to battle."

Why did he go back? Devotion to duty, machismo, hatred of the limelight, love for his fellow Marines, the answer is as complex as the man himself, and as simple as the caption on his arm—*Death Before Dishonor.*

Word of his death spread through the ranks like wildfire, but there was no time to grieve. The battle raged on. Victory on Iwo Jima would take five weeks and cost seven thousand Marine lives. Basilone was wrapped in his rain poncho and buried on the island with the others in a makeshift cemetery under a small wooden cross. Not until three years later were his remains shipped home and laid to rest at Arlington National Cemetery in Virginia with full military honors.

A recommendation for a second Medal of Honor was denied. Instead, his valor on Iwo Jima won him the Navy's second-highest decoration, the Navy Cross. Had the recommendation been granted, he would have been the only serviceman in World War II to win the Medal of Honor twice.

Both of John Basilone's brothers who served in the armed forces during the war, Alphonse, who served in a tank unit under General Patton, and George, a fellow Marine who also fought on Iwo Jima, survived the war and returned to civilian life.

Lena Riggi Basilone left the Marine Corps and started a new life in Los Angeles where she worked at an electric company, immersed herself in veterans groups, and found solace volunteering at a veterans hospital. Her friends recalled a vivacious woman who loved to cook.

She never remarried, telling her niece, "Great love happens only once." Lena passed away in 1999 at age eighty-six and was buried wearing her wedding ring.

Today, a U.S. Navy destroyer is named for John Basilone, Basilone Road runs through Camp Pendleton, and Basilone's picture graces a postage stamp.

His memory endures most vividly in Raritan, New Jersey, home to Basilone Field and the John Basilone Veterans Memorial Bridge. Near the site of his boyhood home is a bronze statue of him high on a pedestal, standing tall as he cradles a Browning machine gun.

Since the 1980s, the town has hosted an annual parade in his honor, a tradition that started with a letter-writing campaign by a third-grade class at a Raritan grammar school—a bunch of kids eager to honor their hero.

John would have liked that.

NOTES ON CHAPTER 1:
"IF THERE IS ONLY ONE PLANE LEFT"

There is an important part of the story of the *Hornet*'s torpedo-plane squadron at Midway that I omitted in order to stay focused on the carrier-based TBDs. A six-plane section of the squadron, which had stayed behind in Norfolk to take delivery of the new Avenger torpedo planes, became part of the land-based air group flying out of Midway that attacked the Kidō Butai on the morning of June 4. Five of the six planes were shot down and all but two of the pilots and gunners were killed.

Fortunately, their compelling tale is included in the highly acclaimed book *A Dawn Like Thunder* (Little, Brown and Company, 2008) by former Congressman Robert Mrazek, which chronicles the service of the *Hornet*'s entire TBD squadron. The book contains a detailed account of the infamous "Flight to Nowhere," led by Stanhope Ring, and the ensuing alleged cover-up of the fiasco by Ring and Admiral Marc Mitscher (see pp. 127–128, 134–135, and Appendix One).

Other book-length histories that provided source material for my book included *Incredible Victory* by Walter Lord (HarperCollins,

NOTES ON CHAPTER 1

1967); *Miracle at Midway* by Gordon Prange, with Donald Goldstein and Katherine Dillon (Penguin Books, 1982); *No Higher Honor: The USS Yorktown and the Battle of Midway* by Jeff Nesmith (Longstreet, 1999); *The Big E: The Story of the USS Enterprise* by Edward Stafford (Bluejacket Books, 2002); *The Unknown Battle of Midway* by Alvin Kernan (Yale University Press, 2005); *Shattered Sword: The Untold Story of the Battle of Midway* by Jonathan Parshall and Anthony Tully (Potomac Books, 2007); *The Battle of Midway* by Craig Symonds (*Oxford University Press*, 2011); and *Enterprise: America's Fightingest Ship and the Men Who Helped Win World War II* by Barrett Tillman (Simon and Schuster, 2012).

The incident involving Lem Massey drinking Scotch in his room the night before the battle is described in Lord, *Incredible Victory*, 84.

Much of the material on Ensign George Gay, including his liaison with the young woman on the train, along with the anecdote about John Waldron firing his .45 into the canebrake, came from Gay's memoir, *Sole Survivor* (Midway Publishers, 1980, 17, 56).

Another useful memoir was *Carrier Combat* by Lieutenant Frederick Mears (Doubleday, Doran and Company, 1944), a member of the *Hornet*'s TBD squadron, who didn't fly with the squadron at the Battle of Midway but served with it afterward, and gives detailed descriptions of a pilot's life aboard a carrier and of attacking an enemy ship in a torpedo plane.

The official action reports by *Yorktown* TBD pilots Harry Corl and Wilhelm Esders and *Enterprise* TBD pilot Lt. (j.g.) Robert Laub describing their squadrons' attacks on the Kidō Butai are on the website titled "The Battle of Midway Roundtable" (midway42.org), an "international association of those having a strong interest in the battle." Wildcat pilot John Thach's account of the battle, published in the June 2007 issue of *Naval History Magazine*, may be found on the website of the U.S. Naval Institute.

NOTES ON CHAPTER 1

A two-hour videotaped interview of Lloyd Childers is on the website of the National World War II Museum under its "Digital Collections." His act of patting the tail of his TBD as it sank is described in Alvin Kernan's book on page 102. Childers's remark on his ninety-third birthday about the battle is from a June 13, 2014, article titled "Oklahoma City Native Reflects on the Battle of Midway 72 Years Later," by Matt Patterson, published in *The Oklahoman*.

Dusty Kleiss's memories of his friend Tom Eversole are from a May 2014 article by Farid Rushdi titled "Pocatello Man Was Lost in the 1942 Battle of Midway," published in the *Idaho State Journal*.

A good source on the *Enterprise*'s TBD squadron at Midway is the serialized diary excerpts of one of its surviving pilots, Lieutenant Irvin McPherson, published in the Sunday edition of the *Chicago Tribune* from November 1942 to January 1943 under the title "I Flew for the Navy."

Some of the background information on John Waldron came from a six-page letter his daughter, Nancy Waldron LeDew, wrote to the South Dakota Hall of Fame in 1985 upon her father's induction. Many thanks to Lori Platzer of that organization for sending me a copy of the letter.

Many thanks also to Adam Minakowski, a librarian at the U.S. Naval Academy's Nimitz Library, for sending me a dossier on Eugene Lindsey pertaining to his time at the academy.

NOTES ON CHAPTER 2:
"THE MOST HATED MAN IN AMERICA"

Two outstanding biographies provided the main source material for this chapter: *Titan: The Life of John D. Rockefeller* by Ron Chernow (Random House, 1998), and the four-volume biography by Allan Nevins, published as *John D. Rockefeller: The Heroic Age of American Enterprise* and *Study in Power: John D. Rockefeller, Industrialist and Philanthropist* (Charles Scribner's Sons, 1940 and 1953 respectively).

Chernow's book tends to be more critical of Rockefeller, Nevins's books more favorable toward him, a difference that is especially noticeable in the way the authors deal with allegations that Rockefeller committed bribery.

According to Chernow, "Rockefeller's papers reveal that he and Standard Oil entered willingly into a staggering amount of [political] corruption," in other words, "paying bribes" (Chernow, *Titan*, 209). Nevins on the other hand found "no concrete evidence of wrongful action ... in the Rockefeller papers" (Nevins, *Study in Power*, vol. 2, 472).

NOTES ON CHAPTER 2

Nevins's other research revealed "a strong suspicion" that an unidentified Standard Oil lobbyist bribed members of the Pennsylvania legislature (Nevins, *Study in Power*, vol. 2, 474). In a conglomerate as massive and as intricate as Standard Oil, however, assuming the bribery occurred, it isn't clear that Rockefeller ordered it or even knew about it.

To support his bribery allegation, Chernow cites two letters to Rockefeller from a New York legislator proposing to defeat legislation harmful to Standard Oil by hiring "attorneys . . . in the Senate [and] Assembly" (Chernow, *Titan*, 210). There is no proof that Rockefeller approved the scheme, or that it was ever carried out.

Chernow claims Standard Oil "bought an exclusive pipeline charter in the Maryland Legislature" that kept a rival pipeline from being built (Chernow, *Titan*, 207). Nevins, however, claims the rival company opted not to build the pipeline in Maryland because it was cheaper to build it in Pennsylvania (Nevins, *Heroic Age*, vol. 1, 576.)

Chernow repeats an allegation by the attorney general of Ohio, Frank Monnett, that Standard Oil offered him a bribe to drop an antitrust case (Chernow, *Titan*, 428). Nevins examined the allegation and concluded that Standard Oil was "certainly not guilty" of trying to bribe Monnett (Nevins, *Heroic Age*, vol. 2, 352–353).

Since, for these reasons, as much as I respect Chernow's work, I question his bribery claim, I decided to omit it from my main text.

Other books I relied on, listed alphabetically by author, included *The Rockefeller Billions: The Story of the World's Most Stupendous Fortune* by Jules Abels (MacMillan, 1965); *The Myth of Antitrust: Economic Theory and Legal Cases* by D.T. Armentano (Arlington House, 1972); *Neighbor John: Intimate Glimpses of John D. Rockefeller in Ormond Beach* by Curt E. Englebrecht (The Telegraph Press, 1936); *The Myth of the Robber Barons*, chap. V, "John D. Rockefeller and the Oil Industry,"

NOTES ON CHAPTER 2

by Burton W. Folsom, Jr. (Young America's Foundation, 1987); *A Rockefeller Family Portrait: From John D. to Nelson* by William Manchester (Little Brown and Company, 1959); *Pocantico: Fifty Years on the Rockefeller Domain* by Tom Pyle as told to Beth Day (Duell, Sloan and Pearce, 1964); *Random Reminiscences of Men and Events* by John D. Rockefeller (Doubleday, Page & Company, 1909); *The History of the Standard Oil Company* by Ida M. Tarbell (two vols.; McClure, Phillips & Co., 1905), and *The American Petroleum Industry: The Age of Illumination, 1859-1899* by Harold F. Williamson and Arnold R. Daum (Northwestern University Press, 1959).

Articles relied on included *The Founding Grandfather* by William Manchester (*New York Times*, October 6, 1974), and *John D. Rockefeller: A Character Study* by Ida Tarbell (*McClure's Magazine*, July and August 1905).

NOTES ON CHAPTER 3:
"RAYS OF GLORY"

Two exceptional books provided most of the material for this chapter: *Washington's Crossing* by David Hackett Fischer (Oxford University Press, 2004), which won a Pulitzer Prize, and *The Winter Soldiers: The Battles for Trenton and Princeton* by Richard M. Ketchum (Henry Holt and Company, 1973). Fischer's work is more scholarly, Ketchum's more dramatic.

Other sources included *1776* by David McCullough (Simon & Schuster, 2005), and *The Battles of Trenton and Princeton* by William S. Stryker (Houghton Mifflin, 1898). The latter work, whose author grew up near Trenton and talked with eyewitnesses as a boy, is the nineteenth century's leading account of the battles. The appendix contains the report of a Hessian court of inquiry and other original documents.

Sergeant Joseph White's narrative is available on the website of *American Heritage* magazine: "The Good Soldier White," *American Heritage* 7, no. 4 (1956).

NOTES ON CHAPTER 3

Private Greenwood's account is in his memoir: *The Revolutionary Services of John Greenwood of Boston and New York*, available at archive.org (see pp. 38–39).

The eyewitness account of Washington's plea to his troops to remain past their enlistments, written by "Sergeant R"—later identified as Sergeant Nathaniel Root of the Twentieth Continental Regiment from Connecticut—is available at jstor.org. (See also Fisher, *Washington's Crossing*, 272–273, and Ketchum, *Winter Soldiers*, 276–279.)

I made an educated guess regarding the unread note to Colonel Rall, which I describe as a "warning that Washington's army had crossed the Delaware and was marching on Trenton." As far as I know, the note wasn't preserved and its exact contents have never been disclosed. I based my surmise about the contents on the fact that Ketchum describes the note as a "warning," and the fact that, upon being shown the note as he lay dying, Rall said, "If I had read this at Mr. Hunt's, I would not be here," a remark that implies that the note, had Rall read it, would have alerted him to Washington's attack (Ketchum, *Winter Soldiers*, 270).

NOTES ON CHAPTER 4:
"THE PERFECT STICK"

Biographies of Coolidge relied on for this chapter include *Calvin Coolidge: The Man from Vermont* by Claude M. Fuess (Little, Brown and Company, 1940); *Coolidge: An American Enigma* by Robert Sobel (Regnery Publishing, 1998); *Calvin Coolidge* by David Greenberg (Henry Holt and Company, 2006); and *Coolidge* by Amity Schlaes (Harper Collins, 2013). The Fuess, Sobel, and Schlaes biographies tend to treat Coolidge more favorably, and the Greenberg biography less so.

Another source was Coolidge's autobiography, which is available at archive.org.

Most of the behind-the-scenes material from Coolidge's White House years came from the memoirs of two members of the White House staff: Edmund Starling, Coolidge's friend and the head of White House security, and Mary Randolph, Grace Coolidge's secretary. The two memoirs are *Starling of the White House: The Story of the Man Whose Secret Service Detail Guarded Five Presidents from Woodrow Wilson to Franklin Roosevelt* (Simon and Schuster, 1946) and *Presidents and First Ladies* (D. Appleton Century Company, 1936).

NOTES ON CHAPTER 4

Many of the facts about American life in the 1920s came from *The Great Crusade and After, 1914–1928* by Preston W. Slosson (Macmillan, 1930), chapters III and VI. The fact that by 1929 America was producing 40 percent (more precisely 42 percent) of the world's goods is from an article titled "Bush's New Axis of Evil" by Patrick J. Buchanan, published in *Human Events* in February 2011.

The claim that Coolidge "slept away" most of his presidency is from Pulitzer Prize winning historian Irwin Unger. The quote appears on page three of Sobel's book.

The remark H. L. Mencken overheard at the 1920 GOP convention about Coolidge's luck is from *A Mencken Chrestomathy*, edited and annotated by Mencken himself (Alfred A. Knopf, 1949), at pp. 252–253. Mencken quotes the remark as follows: "I know Calvin Coolidge inside and out. He is the luckiest goddam _____ in the whole world." I took the liberty of substituting "son of a bitch" for "goddam _____."

The foremost website devoted to Coolidge is that of the "Calvin Coolidge Presidential Foundation" at coolidgefoundation.org.

A film clip of Coolidge giving a short, audible speech on the White House lawn, reportedly the first sound film ever made of a president, is on YouTube.

NOTES ON CHAPTER 5:
"SUZIE Q"

Much of the source material for this chapter came from two biographies of Marciano: *Rocky Marciano: The Rock of His Times* by Russell Sullivan (University of Illinois Press, 2002), which *Booklist* called the "definitive Marciano biography," and *Rocky Marciano, Biography of a First Son* by Everett Skehan (Houghton Mifflin Company, 1977), which according to the author was written with the "total cooperation" of Marciano's brothers, Louis and Peter.

Another important source was Marciano's mini-autobiography, a six-part feature that appeared in the *Saturday Evening Post* between September 15, 1956, and October 20, 1956, which related the ex-champ's life story "as told to" journalists Milton Gross and Al Hirshberg. With varying degrees of candor, Marciano discussed, among other topics, his home life, his relationship with Al Weill, the Vingo knockout, and the Henry Lester fiasco. Marciano claimed kneeing Lester was an accident that happened when Marciano stumbled (see part II, "They Said I'd Get Murdered," *Saturday Evening Post*, September 22, 1956). Skehan, however, quotes Marciano as telling his brother, "Sure I kicked that guy. What was I gonna do, let him beat me?" (p. 50.)

NOTES ON CHAPTER 5

The retired champ's darker side—his friendship with gangsters, his penny-pinching, his womanizing, etc.—is detailed in a controversial hit piece by William Nack in the August 23, 1993, issue of *Sports Illustrated*, which left Marciano's "relatives fuming and [Nack's] sources rushing to claim they were misquoted."

Rocky's brother Peter told the *Boston Globe*, "Rocky may not have been a saint, but this is totally distorted. My brother was a beautiful man, no matter what *Sports Illustrated* says." Rocky's friend Richie Paterniti, whom Nack had interviewed, claimed "Nack distorted every single thing I said. . . . If I said all those things he put in the article, I was lying."

Nack stood by his work, claiming his interview tapes and notes backed up the article. "I just followed my reporting," he said. "I wanted to capture what Rocky was like. Well, this is it." (See "Marciano Story Called a Cheap Shot," *Boston Globe*, August 28, 1993, and "Marciano's Family Mad at *Sports Illustrated*," *Spartanburg Herald-Journal*, August 29, 1993.)

How close Marciano's second fight with Ezzard Charles came to being stopped because of Marciano's cut nostril is a matter of debate. The ringside physician, Dr. Alexander Schiff, was quoted as saying as bad as the cut was, it wouldn't have caused him to stop the fight (Sullivan, *The Rock of His Times*, 231). Reporter Jerry Nason, however, covering the fight at ringside and sitting close enough that some of Marciano's blood stained his shirt, wrote, "At the end of the [seventh] round [Referee Al] Berl said, 'Rocky that's a terrible cut. I'll let the fight go on another round, but then I'll have to stop it'" ("Rocky's Best: 2d Charles Bout," *Boston Evening Globe*, September 3, 1969). Marciano and Charles both claimed the fight was in danger of being stopped (Sullivan, *The Rock of His Times*, 230).

A description of Joe Louis kissing Marciano's casket is on page 256 of Joe *Louis: The Great Black Hope*, by Richard Bak (DA Capo Press, 1998). A picture of him kissing the casket, snapped by UPI photographer Gene Hyde, appeared on page 35 of the September 25, 1969, issue of Jet magazine.

NOTES ON CHAPTER 6:
"HEROES OF THE LAKES"

The main sources for Perry and the Battle of Lake Erie were three biographies of him, one from the nineteenth century and two from the twentieth: *The Life of Commodore Oliver Hazard Perry* by Alex Slidell Mackenzie (Harper and Bros., 1840), available online at archive.org; *We Have Met the Enemy–Oliver Hazard Perry: Wilderness Commodore* by Richard Dillon (McGraw Hill, 1978), and *Oliver Hazard Perry: Honor, Courage, and Patriotism in the Early U.S. Navy* by David Curtis Skaggs (Naval Institute Press, 2006). The book by Mackenzie, whose sister was married to Oliver's brother, is unabashedly hagiographic, but still valuable because the author knew Oliver personally and interviewed many of his family and friends.

Eyewitness accounts of the battle are in *The Travels and Adventures of David C. Bunnell* (J.H. Bortles, 1831) and *Battle of Lake Erie* by Usher Parsons (printed by Benjamin T. Albro, 1853), both of which are available at archive.org.

The main sources for Macdonough and the Battle of Lake Champlain were *The Life of Commodore Thomas Macdonough, U.S. Navy* by his grandson, Rodney Macdonough (The Fort Hill Press, 1909) available

NOTES ON CHAPTER 6

at archive.org; *Thomas Macdonough: Master of Command in the Early U.S. Navy* by David Curtis Skaggs (Naval Institute Press, 2003); and *The Final Invasion: Plattsburgh, the War of 1812's Most Decisive Battle* by David G. Fitz-Enz (Cooper Square Press, 2001).

Other sources included *1812, The Navy's War* by George C. Daughan (Basic Books, 2011); *The Naval War of 1812* by Theodore Roosevelt (G.P. Putnam's Sons, 1902); "The Battle of Lake Erie" by Richard F. Snow (American Heritage Magazine, vol. 27, no. 2, February, 1976); "The Victory on Lake Champlain" by C. S. Forester (American Heritage Magazine, vol. 15, no. 1, December, 1963); and "The Battle of Lake Erie: Building a Fleet in the Wilderness" by Rear Admiral Denys W. Knoll (Naval Historical Foundation, 1979).

A good source on Perry's "Don't Give Up the Ship" flag is "A Flag Bears Witness—Don't Give Up the Ship" by Paulette Dininny (Pennsylvania Heritage Magazine, vol. 38, no. 4, Fall, 2012). No one knows for sure what the flag's original color was: blue, black, brown, or, according to one account, maybe even red. Its current color, brown, may be the result of age. The replica flag atop the rebuilt USS *Niagara* is blue, which is the color I went with.

Skaggs's biography of Perry contains a detailed account of the three-decade controversy over the failure of his second-in-command, Jesse Duncan Elliot, to break formation in the *Niagara* and aid the *Lawrence* as she was being mauled. Elliott's defenders claimed Perry's instructions to his ship commanders to maintain a tight battle line excused Elliot's conduct. The claim is unconvincing. Perry did indeed give such an order, but he had also ordered Elliot to engage the *Queen Charlotte* at close range, an order Elliot ignored when he held the *Niagara* back and fought from a distance. Moreover, regardless of the orders, under the doctrine that no battle plan survives the opening shots, once Elliot saw that the *Lawrence* was in trouble, he was duty bound to sail ahead and support his sister ship (Skaggs, Honor, Courage, and Patriotism, 148).

NOTES ON CHAPTER 7:
"GOD BLESS AMERICA"

The main source of information about Irving Berlin's life was the biography *As Thousands Cheer: The Life of Irving Berlin* by Laurence Bergreen (Viking Penguin, 1990). Other Berlin biographies that provided source material included: *The Story of Irving Berlin* by Alexander Woollcott (G.P. Putnam's Sons, 1925); *Irving Berlin* by Michael Freedland (Stein and Day, 1974); *Irving Berlin: A Life in Song* by Philip Furia (Schirmer Trade Books, 1998); and *Irving Berlin: New York Genius* by James Kaplan (Yale University Press, 2019).

A widely held belief, repeated in Berlin's New York Times obituary, that he "never learned to read or write music," is a myth debunked by a handwritten sheet of music containing this note in his handwriting: "1st lead sheet ever taken down by Irving Berlin August 16, 1932." (see Kaplan, *A New York Genius*, 75 and 344, fn. 4; see also *The Irving Berlin Reader*, Benjamin Sears, ed., Oxford University Press, 2012, pp. 200–201.) Although Berlin apparently taught himself the rudiments of reading and writing music, he would always use a transcriber to put the final version of a song on paper (Bergreen, *As Thousands Cheer*, 585).

NOTES ON CHAPTER 7

A good source on "God Bless America" is a book that deals wholly with that song: *God Bless America: The Surprising History of an Iconic Song* by Sheryl Kaskowitz (Oxford University Press, 2013). The book *This Land that I Love: Irving Berlin, Woody Guthrie and the Story of Two American Anthems* by John Shaw (Public Affairs, 2013) chronicles "God Bless America," as well as the classic song written in rebuttal, Guthrie's "This Land Is Your Land."

A good source for what life was like in Manhattan's Lower East Side at the time of Berlin's youth is *At the Edge of a Dream: The Story of the Jewish Immigrants on New York's Lower East Side, 1880–1920* by Lawrence Epstein (Jossey-Bass, 2007).

The main sources on Kate Smith's life were her autobiography, *Upon My Lips a Song* (Funk and Wagnalls, 1960), and the biography *Kate Smith* by Richard Hayes (McFarland & Company, Inc., 1995). Printed excerpts from Smith's daytime radio show may be found in *Kate Smith Speaks—50 Selected Original Radio Scripts: 1938–1951* edited by Richard Hayes (BearManor Media, 2013). A recording of Smith's original 1938 performance of "God Bless America," as well as a video of her last performance of the song at the Philadelphia Flyers' game in 1976, may be found on YouTube.

Sadly, in 2019, after honoring Smith's memory for decades, the Philadelphia Flyers' management ordered her statue outside the team's arena removed because nearly ninety years earlier, in the early 1930s, she had performed two songs that were allegedly racist: "That's Why Darkies Were Born" and "Pickaninny Heaven."

Neither song, however, was generally considered racist at the time she sang it. Paul Robeson, the black singer and civil rights activist, also recorded "That's Why Darkies Were Born." Songs about "Pickaninnies" were common and sometimes recorded by black artists. Robeson recorded a song called "Pickaninny Slumber Song." Ethel Waters, another black singer, recorded "Pickaninny Blues."

NOTES ON CHAPTER 8:
"THE YEGG HUNTERS"

Sources for this chapter included *Public Enemies* by Bryan Burrough (Penguin Press, 2004); *The Vendetta: FBI Hero Melvin Purvis's War against Crime, and J. Edgar Hoover's War against Him* by Alston Purvis (Melvin Purvis's son) and Alex Tresniowski (Public Affairs, 2005); *Baby Face Nelson: Portrait of a Public Enemy* by Steven Nickel and William J. Helmer (Cumberland House Publishing, Inc., 2020); *The Life and Death of Pretty Boy Floyd* by Jeffery S. King (The Kent State University Press, 1998); and *Killing The Mob* by Bill O'Reilly and Martin Dugard (St. Martin's Press, 2021).

A good website for information on the early FBI and its agents is *Faded Glory: Dusty Roads of an FBI Era*, at historicalgmen.squarespace.com, which includes thumbnail biographies of the agents in the Dillinger Squad.

Forty-five years after Pretty Boy Floyd's death, a lawman who witnessed it, Officer Chester Smith, of the East Liverpool, Ohio, police, made headlines by claiming that, as Floyd lay wounded but very much alive, an FBI agent walked up and executed him. In his

NOTES ON CHAPTER 8

book *Public Enemies*, however, author Bryan Burrough convincingly discredits Smith's claim (see footnote, p. 468).

There is some confusion about when the Federal Bureau of Investigation got its name. What had been the "Bureau of Investigation" wasn't officially designated the "Federal Bureau of Investigation" until 1935, when the War on Crime was more than half over. Unofficially, however, the bureau was known as the Federal Bureau of Investigation as early as August 1933, just a month after FDR's attorney general, Homer Cummings, publicly announced the war. (See "Bill Numbers Listed in Urschel Inquiry," *New York Times*, August 3, 1933, p. 4.)

NOTES ON CHAPTER 9:
"I AIN'T NO MUSEUM PIECE"

Full-length book sources for this chapter included *Hero of the Pacific: The Life of Marine Legend John Basilone* by James Brady (John Wiley & Sons, 2010); *Red Blood, Black Sand: Fighting Alongside John Basilone From Boot Camp to Iwo Jima* by Chuck Tatum (Penguin Group, 2012); *The Battle for Guadalcanal* by Samuel Griffith II (Nautical and Aviation Publishing Company of America, 1963); *Guadalcanal* by Richard Frank (Random House, 1990); *Challenge for the Pacific, Guadalcanal: The Turning Point of the War* by Robert Leckie (Bantam Books, 1965); *Guadalcanal, Starvation Island* by Eric Hammel (Pacifica Military History, 1987); and *The Battle for Iwo Jima, 1945* by Derrick Wright (The History Press, 1999).

Raritan's Hero, The John Basilone Story by Bruce Doorly, a medium-length biography, is available on the internet at *raritan-online.com*.

Sources also included two newspaper serials on John Basilone's life: "Shooting Star: The Story of WWII Hero John Basilone" by Keith Sharon, which appeared in the *Orange County* [California] *Register* in six installments from September 19, 2004 to September 24,

NOTES ON CHAPTER 9

2004, and "The Basilone Story" by Phyllis Basilone Cutter, the hero's sister, which appeared in fourteen installments in the *Somerset* [New Jersey] *Messenger Gazette* from November 15, 1962 to February 14, 1963. Although Phyllis Cutter's account contains obvious errors (see, e.g., Brady, Hero of the Pacific, 206), it's still a valuable resource.

Many thanks to Ken Kaufman of the Bridgewater branch of the Somerset County, New Jersey, library system for sending me all fourteen parts of Ms. Cutter's series.

The Raritan, New Jersey, city library has "Basilone Archives" on its website, which include many photos of the hero, along with firsthand sources, most notably a four-page report—neatly typed, lucid, and very matter-of-fact—written by Basilone himself on the action that won him the Medal of Honor.

It has often been claimed that he was the first enlisted Marine in World War II to win the Medal of Honor. In fact, that distinction belongs to Sergeant Clyde Thomason of the 2nd Raider Battalion, who won the medal for valor during a raid on Makin Island in August 1942 (Brady, *Hero of the Pacific*, 9). Sergeant Thomason was killed during the raid, making Basilone the first enlisted Marine to win the Medal of Honor and survive.

Although the medal is officially called the "Medal of Honor," I also called it the "Congressional Medal of Honor" because many people know it by that name.

ACKNOWLEDGEMENTS

Many of the images—photos, paintings, etc.— reproduced in this book are either in the public domain or appear via licenses purchased from vendors such as Getty Images and Shutterstock, or from the Associated Press.

Many other images, however, were made available through the generosity of curators, archivists, photographers, librarians, and family members, who, free of charge or for a nominal fee, shared works under their control.

Thus, in the order the images appear, heartfelt thanks are due to:

Steven Childers, nephew of Lieutenant Colonel Lloyd Childers; Michele Hiltzik Beckerman, Assistant Director for Reference, Rockefeller Archive Center; Eleanor Gillers, Head of Rights and Reproductions, New York Historical Society Museum and Library; Carolyn Cruthirds, Coordinator of Image Licensing, Boston Museum of Fine Arts; Michael P. Miller, Assistant Head of Manuscripts Processing and Registrar, Library and Museum, American Philosophical

ACKNOWLEDGEMENTS

Society; Julie Bartlett Nelson, Archivist, Calvin Coolidge Presidential Library and Museum, Hampshire Room for Local History, Forbes Library; Juls Sundberg, Cataloger and Metadata Librarian, Vermont Historical Society; Laura V. Trieschmann, State Historic Preservation Officer, Vermont Division of Historic Preservation; Nicole Casper, Director of Archives, Stonehill College; Allen C. Browne, photographer; Pam Overmann, Curator, Navy Art Collection; Bill Teaney, Business Project Manager, Naval Academy Business Services Division; Joan Thomas, Art Curator, National Museum of the Marine Corps.

By request, the acknowledgement for the use of the lyrics of the song "God Bless America" is as follows:

God Bless America®
Words and Music by Irving Berlin
© Copyright 1938, 1939 by Irving Berlin
Copyright Renewed 1965, 1966 by Irving Berlin
Copyright Assigned to the Trustees of the God Bless America
Reprinted by Permission of Hal Leonard LLC

www.ingramcontent.com/pod-product-compliance
Lightning Source LLC
Chambersburg PA
CBHW030231170426
43201CB00006B/181